International Assessment of
Research and Development in
Catalysis by Nanostructured Materials

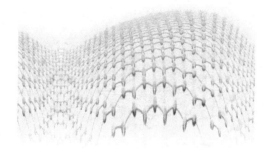

International Assessment of Research and Development in
Catalysis by Nanostructured Materials

Editor
Robert Davis
University of Virginia, USA

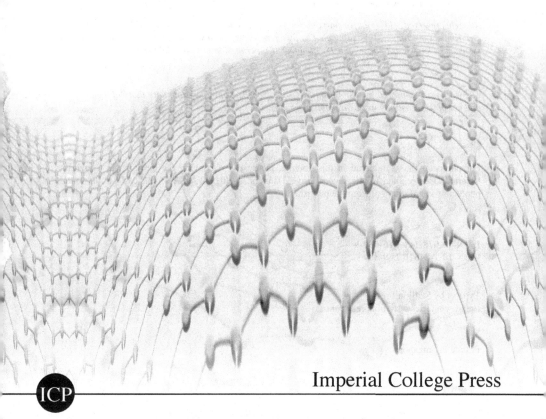

Imperial College Press

Published by

Imperial College Press
57 Shelton Street
Covent Garden
London WC2H 9HE

Distributed by

World Scientific Publishing Co. Pte. Ltd.
5 Toh Tuck Link, Singapore 596224
USA office: 27 Warren Street, Suite 401-402, Hackensack, NJ 07601
UK office: 57 Shelton Street, Covent Garden, London WC2H 9HE

British Library Cataloguing-in-Publication Data
A catalogue record for this book is available from the British Library.

INTERNATIONAL ASSESSMENT OF RESEARCH AND DEVELOPMENT IN CATALYSIS BY NANOSTRUCTURED MATERIALS
Copyright © 2011 by Imperial College Press

All rights reserved. This book, or parts thereof, may not be reproduced in any form or by any means, electronic or mechanical, including photocopying, recording or any information storage and retrieval system now known or to be invented, without written permission from the Publisher.

The work was sponsored by NSF Grant ENG-0739505. The opinions stated are those of the authors, and not of their institutions or the sponsoring agencies.

For photocopying of material in this volume, please pay a copying fee through the Copyright Clearance Center, Inc., 222 Rosewood Drive, Danvers, MA 01923, USA. In this case permission to photocopy is not required from the publisher.

ISBN-13 978-1-84816-689-9
ISBN-10 1-84816-689-3

Typeset by Stallion Press
Email: enquiries@stallionpress.com

Printed in Singapore.

Dedication

We at WTEC wish to extend our gratitude and appreciation to the panelists for their valuable insights and their dedicated work in conducting this international benchmarking study of R&D in catalysis by nanostructured materials. We wish also to extend our sincere appreciation to the presenters at the North American baseline workshop and to the panel's site visit hosts for so generously and graciously sharing their time, expertise, and facilities with us. For their sponsorship of this important study, our thanks go to the National Science Foundation (NSF), the Department of Energy (DOE), the Air Force Office of Scientific Research (AFOSR), and the Defense Threat Reduction Agency (DTRA). We believe this report provides a valuable overview of ongoing R&D efforts in catalysis by nanostructured materials that can help scientists and policymakers effectively plan and coordinate future efforts in this important field.

R. D. Shelton

Contents

Foreword   xiii

Abstract   xv

Executive Summary   xvii

1. Overview of Catalysis by Nanostructured Materials
   Robert J. Davis

1.1 Introduction to the Study	1
1.2 Approach and Methodologies	5
1.3 Report Structure	8
1.3.1 Bibliometric analysis	8
1.4 Investment Models and Trends	12
1.4.1 The United States	12
1.4.2 Asia	13
1.4.3 Western Europe	15
1.4.4 Summary	18
1.5 General Observations	20
1.6 Technical Themes of the Study	22
1.7 Conclusions	23
References	24

2. Synthesis of Nanostructured Catalysts
   Raul F. Lobo

2.1	Introduction	25
2.2	Rhenium Clusters in Zeolite ZSM-5	28
2.3	Novel Propene Partial Oxidation Catalysts	30
2.4	Mesoporosity Designed into Microporous Catalysts	31
	2.4.1 Micro-mesoporous zeolites by design of organic-inorganic surfactants	32
	2.4.2 Micro-mesoporous zeolites from carbon templates	34
	2.4.3 Micro-mesoporous catalysts by assembly of nanoparticles	35
2.5	Synthesis of Extra-Large-Pore Zeolite ITQ-33	36
2.6	Heteropolyanions as Precursors for Desulfurization Catalysts	39
2.7	Final Remarks	40
References		40

3. Spectroscopic Characterization of Nanostructured Catalysts
   Jeffrey T. Miller

3.1	Background	43
3.2	Laboratory Characterization Methods	44
	3.2.1 Overview	44
	3.2.2 New spectroscopic capabilities and adaptations to standard laboratory instruments	45
3.3	Synchrotron Methods	52
	3.3.1 Overview	52
	3.3.2 Scattering techniques	53
	3.3.3 Millibar XPS at BESSY synchrotron (Germany)	58
3.4	Conclusions	60
References		61

4. Electron and Tunneling Microscopy of Nanostructured Catalysts
   Renu Sharma

4.1	Introduction	65

		4.1.1	Overview of high resolution characterization techniques	66
	4.2		General Characterization of Catalyst Particles	70
	4.3		Nanostructure Characterization Under Working Conditions	74
		4.3.1	Effect of environment on surface structure and reactivity	75
	4.4		Future Trends	85
	4.5		Summary	87
	References			87

5. Theory and Simulation in Catalysis
   Matthew Neurock

	5.1	Introduction	91
	5.2	Computational Catalysis: Where Are We Today?	94
	5.3	Methods and Their Applications	94
		5.3.1 Electronic structure methods	94
		5.3.2 Atomic and molecular simulations	98
		5.3.3 Dynamics	99
		5.3.4 Kinetics	100
	5.4	Snapshot of the Efforts in Europe and Asia	101
		5.4.1 Europe	101
		5.4.2 Asia	106
		5.4.3 Comparison of Europe, Asia, and the United States	107
	5.5	Universal Trends	110
		5.5.1 The good news	110
		5.5.2 The not-so-good news	112
	5.6	Examples of Applications of Theory and Simulation	114
		5.6.1 Connecting theory and spectroscopy	114
		5.6.2 Modeling more realistic reaction environments	121
		5.6.3 Applications to energy	129
		5.6.4 Simulating catalytic performance	131
		5.6.5 Design in heterogeneous catalysis	135
		5.6.6 From theory to synthesis	138

5.7	Summary and Future Directions	140
References		143

6. Applications: Energy from Fossil Resources
Levi Thompson

6.1	Introduction	151
6.2	Production of Liquid Fuels	153
	6.2.1 Catalysts for petroleum refining	153
	6.2.2 Catalysts for syngas conversion	158
6.3	Production of Hydrogen	160
	6.3.1 University of Udine, Consiglio Nazionale delle Ricerche (Italy)	163
	6.3.2 University of Trieste, Consiglio Nazionale delle Ricerche (Italy)	164
	6.3.3 Instituto di Chimica dei Composti OrganoMetallici (Italy)	164
	6.3.4 Tokyo Metropolitan University Department of Applied Chemistry	165
	6.3.5 Tsinghua University (China)	165
	6.3.6 Tianjin University (China)	166
6.4	Fuel Cell Research	167
	6.4.1 University of Trieste, Consiglio Nazionale delle Ricerche (Italy)	167
	6.4.2 Tsinghua University (China)	168
	6.4.3 Tianjin University (China)	168
6.5	Environmental Catalysis	169
6.6	Three-Way Catalysis	170
	6.6.1 Toyota Motor Corporation, Higashi-Fuji Technical Center	170
	6.6.2 NO_x selective catalytic reduction	170
	6.6.3 CO_2 reduction	172
6.7	Summary	175
	6.7.1 Project highlights: Energy-centered catalysis R&D	175
	6.7.2 Regional characteristics of catalysis R&D for improving fossil energy production	176
References		176

7. Applications: Chemicals from Fossil Resources
   Vadim V. Guliants

   7.1 Introduction 185
   7.2 Alkylation 186
   7.3 Dehydrogenation and Hydrogenation 195
       7.3.1 Dehydrogenation 196
       7.3.2 Hydrogenation 204
   7.4 Selective Oxidation 212
       7.4.1 Selective oxidation catalysis by nanosized gold and other noble metals 213
       7.4.2 Selective oxidation of lower alkanes by bulk mixed metal oxides 218
       7.4.3 Catalytic behavior of Mo-V-(Te-Nb)-O M1 phase catalysts 220
       7.4.4 On cooperation of M1 and M2 phases in propane ammoxidation 221
       7.4.5 Surface termination of M1 phase 222
   7.5 Future Trends 224
   References 229

8. Applications: Renewable Fuels and Chemicals
   George Huber

   8.1 Introduction 239
   8.2 Key Observations 240
   8.3 Biomass Conversion 241
       8.3.1 Biomass feedstocks 241
       8.3.2 Liquid fuels from biomass 244
       8.3.3 Biomass gasification and syngas conversion . . . 245
       8.3.4 Fast pyrolysis and bio-oil upgrading 247
       8.3.5 Liquid-phase/aqueous-phase catalytic processing . 251
       8.3.6 Vegetable oil conversion 252
       8.3.7 Chemicals from biomass 254
       8.3.8 Bibliometric analysis of catalysis and biofuels . . 255
   8.4 Photocatalytic Water Splitting 256
   8.5 Conclusions 258
   References 259

Appendix 1: Panelists' Biographies 263
Appendix 2: Bibliometric Analysis of Catalysis Research, 273
 1996–2005
Appendix 3: Glossary 291

Index 299

Foreword

> We have come to know that our ability to survive and grow as a nation to a very large degree depends upon our scientific progress. Moreover, it is not enough simply to keep abreast of the rest of the world in scientific matters. We must maintain our leadership.[1]

President Harry Truman spoke those words in 1950, in the aftermath of World War II and in the midst of the Cold War. Indeed, the scientific and engineering leadership of the United States and its allies in the twentieth century played key roles in the successful outcomes of both World War II and the Cold War, sparing the world the twin horrors of fascism and totalitarian communism, and fueling the economic prosperity that followed. Today, as the United States and its allies once again find themselves at war, President Truman's words ring as true as they did a half-century ago. The goal set out in the Truman Administration of maintaining leadership in science has remained the policy of the US Government to this day: Dr. John Marburger, the Director of the Office of Science and Technology Policy

---

[1] Remarks by the President on May 10, 1950, on the occasion of the signing of the law that created the National Science Foundation. *Public Papers of the Presidents* 120:338.

(OSTP) in the Executive Office of the President made remarks to that effect during his confirmation hearings in October 2001.[2]

The United States needs metrics for measuring its success in meeting this goal of maintaining leadership in science and technology. That is one of the reasons why the National Science Foundation (NSF) and many other agencies of the US Government have supported the World Technology Evaluation Center (WTEC) and its predecessor programs for the past 20 years. While other programs have attempted to measure the international competitiveness of US research by comparing funding amounts, publication statistics, or patent activity, WTEC has been the most significant public domain effort in the US Government to use peer review to evaluate the status of US efforts in comparison to those abroad. Since 1983, WTEC has conducted over 60 such assessments in a wide variety of fields, from advanced computing, to nanoscience and technology, to biotechnology.

The results have been extremely useful to NSF and other agencies in evaluating ongoing research programs, and in setting objectives for the future. WTEC studies also have been important in establishing new lines of communication and identifying opportunities for cooperation between US researchers and their colleagues abroad, thus helping to accelerate the progress of science and technology generally within the international community. WTEC is an excellent example of cooperation and coordination among the many agencies of the US Government that are involved in funding research and development: almost every WTEC study has been supported by a coalition of agencies with interests related to the particular subject at hand.

As President Truman said over 50 years ago, our very survival depends upon continued leadership in science and technology. WTEC plays a key role in determining whether the United States is meeting that challenge, and in promoting that leadership.

Michael Reischman
Deputy Assistant Director for Engineering
National Science Foundation

---

[2] http://www.ostp.gov./html/01_1012.html.

Abstract

This WTEC panel report assesses the international research and development activities in the field of catalysis by nanostructured materials. Catalysis is important for a wide variety of processes that impact manufacturing, energy conversion, and environmental protection. This study focused specifically on solid catalysts and how nanoscale structures associated with them affect their reactivity. The principal technical areas of the study are (a) design and control of synthetic nanostructures; (b) nanoscale characterization of catalysts in their working state; (c) theory and simulation; and (d) applications. The panel visited over 40 institutions and companies throughout East Asia and Western Europe to explore the active research projects in those institutions, the physical infrastructure used for the projects, the funding schemes that enable the research, and the collaborative interactions among universities, national laboratories, and corporate research centers.

A bibliometric analysis of research in catalysis by nanostructured materials published from 1996 to 2005 was conducted as part of this WTEC study. The total number of published papers as well as the expected total number of citations of those papers revealed a growing focus on this subject. Western Europe was the numerical output leader in the world; US output, while published in high-impact journals, was relatively stagnant, and the number of published papers originating from China was growing exponentially and expected to exceed that from the United States in the latter half of this decade. China's rapidly expanding economy together with its growth in large-scale chemical and refining plants motivate its significant commitment to catalysis research.

The panel found that cooperation between universities and companies in catalysis R&D is common in Europe and Asia, presumably because of a more favorable intellectual property environment outside of the United States. In the area of catalyst synthesis, there is substantial activity to develop microporous materials with controlled mesoporosity and to prepare nanosized particles with preferentially exposed crystal planes. Recent advances in spectroscopy and microscopy allow the nanostructures of catalyst particles to be examined under more realistic environmental conditions approaching those of industrial reactions. Electronic structure methods and molecular simulations are now considered to be necessary tools for use alongside experiments to help guide catalysis research. The applications of much of the research observed by the panel are directly related to energy and the environment.

Executive Summary

Robert J. Davis

The science and technology of catalysis are important in the production of modern medicines, new fibers for clothing and construction, a wide variety of consumer products, cleaner-burning fuels, and environmental protection. The US chemical industry, which relies heavily on catalysis, had record exports in 2007 and its first trade surplus since 2001. Furthermore, catalyst technologies account for over US$1 trillion of revenues in the US economy and about a third of its material GDP. Catalysts are considered the engines that power the world at the nanometer scale and are generally considered to be the most successful application of nanotechnology. For these reasons, the World Technology Evaluation Center initiated a study in 2006 to assess the research and development activities related to catalysis by nanostructured materials under the sponsorship of the National Science Foundation (NSF), the Department of Energy (DOE), the Air Force Office of Scientific Research (AFOSR) and the Defense Threat Reduction Agency (DTRA). The information from the assessment will be used to identify high-impact research areas worth exploring in US R&D programs, clarify research opportunities and needs for promoting progress in the field, identify opportunities and mechanisms for international collaboration, and evaluate the position of foreign research programs relative to those in the United States. The principal technical focus areas of the study were

- design and control of synthetic catalytic structures
- nanoscale characterization of catalysts in their working state
- theory and simulation
- applications.

WTEC recruited a panel of eight US experts in the field to perform the assessment. Although a worldwide assessment of this kind normally begins with a benchmarking exercise to establish the state-of-the-art in US R&D programs, a report from the 2003 US National Science Foundation workshop "Future Directions in Catalysis: Structures that Function at the Nanoscale" (NSF, 2003) served as the reference point for US activities in the focus area of the study. Therefore, the current project was performed in the following four phases:

(1) *Conduct a bibliometric analysis to establish world trends in publishing.* The analysis examined publication trends by various countries over the last decade. Results were used to confirm and identify new sites selected by the panelists to visit.
(2) *Visit a number of the world's leading university, government, and industry laboratories.* The WTEC panelists visited 20 sites in China, Japan, and the Republic of Korea during June of 2007 and 22 sites in Europe in September of 2007.
(3) *Report findings in a public workshop.* The panelists reported their findings to the US sponsors, the catalysis community and the general public at a workshop held at the FDIC L. William Seidman Center in Arlington, VA, on November 29, 2007. The workshop, entitled "Assessment of International R&D in Catalysis by Nano-structured Materials", was a public forum in which panelists presented overall findings, specific examples from the site visits, and general conclusions. The workshop also allowed for discussion and critical review of the findings.
(4) *Compile the findings into a written report.* This document represents the written record of the study that will be made available to the

sponsors, funding agencies, policymakers, the catalysis community, and the general public. This report is available on the Internet at http://www.wtec.org/catalysis/.

Principal Findings

Bibliometric analysis of catalysis papers

Several important quantitative outputs were determined by the bibliometric analysis (see full analysis in Appendix B). First, catalysis papers in the decade of the study, 1996–2005, represented about 1/60th of the world science output, and the growth rate in catalysis papers (5.4%) exceeded that of all science papers (2.9%), indicating a growing importance of catalysis research in the world. The output of catalysis papers by 13 Western European countries exceeded the total produced by the United States by almost a factor of two. Moreover, there was a significant growth in the number of papers from the People's Republic of China from 1996–2005, while the number of papers from the United States over the same time period was essentially stagnant. In 1996, the catalysis papers from the United States outnumbered those from China by a factor of 6; however, the large disparity between China and the United States was largely eliminated by 2005, and reasonable extrapolation of the exponential growth of Chinese output suggests that the United States now lags China in the number of catalysis papers.

A second quantitative aspect of the bibliometric analysis involved a country's relative commitment (RC) to catalysis. The RC measures the importance of catalysis research within a particular country, with the RC value for the entire world being unity. The very high relative commitment of China (RC > 2) to catalysis research over the decade of the WTEC bibliometric study is consistent with its exponential growth in published papers and contrasts with a very low commitment of the United States to catalysis research (RC ~0.6). The relative commitment of Western Europe to catalysis varies by country, but is generally near the world average (RC = 1).

In terms of qualitative comparisons, the papers published by US researchers appear in journals with the highest potential impact, whereas the papers published in East Asia appear in journals with below average impact factors. The countries of Western Europe publish in journals with impact factors near the world average for the field. Since the Western European countries publish a high volume of papers, the total impact from Western Europe dominated the world over the decade of the WTEC bibliometric study, and the gap between Western Europe and the United States widened over that decade.

The results of the WTEC bibliometric analysis suggest that the United States had a significant, but not dominant, position in catalysis research over the decade 1996–2005. Western Europe led the world in both total number of papers published and the total citation impact. However, the very high relative commitment of East Asia to catalysis research, particularly in China, has fueled an exponential growth in the number of published papers in the last decade and will soon challenge the position of both the United States and Western Europe.

Financial support of catalysis research

The financial models used to support catalysis research vary widely around the world. As a baseline for comparison, the US National Science Foundation and Department of Energy support catalysis research at universities at a level of approximately US$30 million per year, with nearly half of that originating from the Basic Energy Sciences program at DOE. The Catalysis and Biocatalysis Program at NSF, together with contributions to special programs such as Nanoscale Interdisciplinary Research Teams (NIRT), Nanoscale Exploratory Research (NER), and Partnerships for International Research and Education (PIRE), contribute approximately US$7 million per year to US university research. The total support from the DOE Office of Science, the DOE Technology Offices, other governmental agencies, and US business for catalysis research at the US national laboratories is estimated to be about US$45 million. Although catalysis research is also funded by other agencies of the US government such as the Department of Defense and by the private sector through corporate contracts, the major research activities in fundamental catalysis are supported by NSF and DOE.

The different financial models used throughout the world all appear to be effective at supporting catalysis research, and no single system can be declared as the best. However, certain aspects of each deserve close examination for possible adoption in the United States. For example, the sustained level of baseline funding provided to a science-oriented institution such as the Fritz Haber Institute (FHI; see Site Reports-Asia on the International Assessment of Research in Catalysis by Nanostructured Materials website, www.wtec.org/private/catalysis; password required, contact WTEC) allows for effective utilization of a highly trained research staff that is necessary to build the world's next-generation scientific instruments. However, a stable funding stream is not the only unique feature that allows the institution to function at a very high level scientifically. Justification of research directions at the FHI is based on recommendations from an external evaluation committee of scientific experts from around the world instead of on classical numerical outputs such as papers and patents. The stable source of funding for long-term research goals at the FHI contrasts with the typical three-year granting cycle at the US government agencies, which also require yearly justifications for grant expenditures. This difference in funding strategy between the US government and the FHI was highlighted during the WTEC visit as a major reason why certain kinds of research and instrument development are unique to the FHI.

The consortia model developed within and throughout the European countries is another strategy for funding research. These consortia tend to be most effective within a single country, presumably because of the ease of communication, shared cultural identity, and geographic proximity of participating institutions. Although pan-European initiatives have also experienced success, a significant fraction of funding in those initiatives is dedicated to travel and collaboration instead of directly supporting research.

European and Asian countries have done an excellent job combining academic research with national laboratory activities. The combination is almost seamless at many institutions, with principal investigators, professors, doctoral students, and research staff members working together on common goals. There appears to be a more significant separation between academic catalysis research in the United States and its complement at the mission-oriented DOE national laboratories.

Support of doctoral students and postdoctoral researchers on US research grants tends to consume the major fraction of direct charges. In countries like China and Korea, government programs that fund students directly for their graduate education in areas of national importance remove the need to fund students directly on research grants. Although it is unclear whether or not that system is superior to funding students directly on grants, it does appear to be similar to the graduate training grant programs in the United States that are common in other fields.

WTEC panelists found significant industrial collaborations at most of the sites and had the impression that industrial support was more prevalent outside of the United States. The reasons appear to be many, such as a lower cost of performing research outside of the United States, a more cooperative intellectual property environment outside of the United States, and possible access to emerging world markets for the next users of catalytic technologies. Moreover, the panel noted in its visits overseas the high quality of pilot plant facilities and catalytic reactor systems capable of industrial operating conditions located within academic settings. The infrastructure for fundamental research on catalyst synthesis and characterization combined with industrially-relevant catalyst testing appeared to exceed that of typical academic catalysis laboratories in the United States.

Routine equipment for catalysis research such gas chromatographs, reactors, vacuum chambers, adsorption systems, and bench-top spectrometers, as well as major research instruments such as electron microscopes, nuclear magnetic resonance spectrometers, and X-ray diffractometers, are generally available in labs worldwide. In addition, researchers appear to have ready access to synchrotron light sources for advanced *in situ* characterization of catalysts. However, the WTEC panelists reported a significant concern with the aging US catalysis infrastructure compared to that observed in East Asia and Europe. The difficulties US researchers currently encounter with acquisition of new instruments routinely used for catalysis research could severely impact US competitiveness in the very near future.

General Observations

Catalysis is often associated with large-scale chemical, petrochemical, or oil refinery processes, which are areas of rapid growth in Asia but are fairly stagnant areas in the United States and the European Union. Although Asia's rapidly expanding economy accounts for its major growth in catalysis research, the reasons for the dominant position of Western Europe in catalysis research output are not so straightforward. Several key factors are likely to contribute to this result. First, catalysis is generally viewed outside of the United States as a fundamental science that enables discovery and development of technology in a variety of energy- and chemicals-related fields. Therefore, catalysis research is pursued in chemistry and physics departments throughout Europe, with much smaller levels of activity in European chemical engineering departments. Within the United States, heterogeneous catalysis has been viewed as a mature field that is studied mainly in chemical engineering departments. Chemistry departments in the United States tend to support much more fundamental studies involving model surfaces in pristine environments, homogeneous or single-site molecular catalysis, and biocatalysis. This artificial division allowed heterogeneous catalysis in the United States to be incorrectly perceived as an applied field of research instead of one with the potential for fundamental discovery.

Another reason European catalysis appears to have a dominant position is the close coupling of universities and national laboratories with industry. Companies within several of the countries the WTEC panel visited appear to campaign for catalysis research at the national level. This kind of advocacy support from companies, together with research contracts involving universities and government labs, indicates a significant role of the private sector in setting research directions.

Moreover, the current intellectual property (IP) environment in Europe appears to foster university-company relations, at least more so than in the United States. However, there are indications that European universities are beginning to explore the position of many US universities on IP ownership and may soon move in the direction of the United States.

Finally, the overall level of investment in catalysis research in Europe simply appears to be higher than that in the United States. Although the United States is starting to see large block grants from companies such as BP and the Dow Chemical for catalysis research at universities, the combined investment of the US government (mainly NSF and DOE) and the industrial sector in university and national laboratory research appears to be far below that of the European countries.

The WTEC panelists observed the use of high-throughput instrumentation in both Asia and Europe. For example, robotic synthesis of zeolite materials and high-throughput reaction testing allowed for discovery of new catalytic materials with unique properties at the Instituto de Tecnología Química (ITQ) in Valencia, Spain (see Site Reports-Europe on the International Assessment of Research in Catalysis by Nanostructured Materials website, www.wtec.org/private/catalysis; password required, contact WTEC). Also, the Center for Microchemical Process Systems at KAIST (Korea Advanced Institute of Science and Technology) makes extensive use of high-throughput screening methods for discovery of new materials. New instrumentation recently purchased at institutions throughout Asia and Europe was aimed at rapid analysis of catalyst samples. Since research abroad is often carried out in major centers of activity, some of the instrumentation was actually invented and constructed in-house. This model contrasts with the operation of most US academic laboratories, which have neither in-house expertise for tool creation nor the resources for tool construction.

The WTEC panel also noted the effective use of permanent research staff positions in laboratories outside of the United States. In the vast majority of sites visited, including those associated with institutes and universities, the number of dedicated staff members in support of the research activities appeared to far exceed those associated with US laboratories. The employment of highly skilled technical staff members at the Fritz Haber Institute was highlighted during the WTEC visit as a major reason why next-generation research instruments can be designed and built within the facility. The instruments were far too complex to be constructed solely by graduate students and short-term postdoctoral researchers. The US current funding models do not support the same level of technicians, and academic institutions do not appear to have funds available for additional positions; in fact, many US researchers are experiencing a decrease in technical staff at their universities.

Technical Themes of the Study

The chapters in this report present the findings of the panel on the technical themes of the study: synthesis, *in situ* characterization, theory and simulation, and applications. Regarding catalyst synthesis, several countries were actively pursuing new methods to produce highly structured solids such as zeolites and carbons having both micropores and mesopores. Moreover, preparation and stabilization of metal and metal oxide nanoparticles with controlled facets, sizes, and compositions are also areas of wide interest. Characterization of nanostructured catalysts by environmental electron microscopy is revealing how nanoscopic features are affected by the surrounding environment; this advanced method of observation will be pursued more aggressively with the increasing availability of commercial instruments. *In situ* spectroscopy has now become a routine method for catalyst characterization around the world. However, rapid improvements in temporal and spatial resolution of many spectroscopic methods, as well as adaptation of methods to allow for interrogation of catalysts under industrial reaction conditions, are enabling collection of an unprecedented level of new information on the structure of catalysts in their working state. Theory and modeling have gained universal acceptance as necessary tools for advancing catalysis science. Improvements in method accuracy, computational speed, and model development have moved theory to a position alongside experimentation in many laboratories. Theory has excelled in the prediction of atomic structure and spectroscopic features of catalytic materials. Although the prediction of reaction kinetics is still developing, theory is being used to suggest novel compositions to improve catalytic performance.

The key applications stimulating most of the catalysis research worldwide were related to energy and the environment. Conversion of nonpetroleum feedstocks such as coal, natural gas, and biomass to energy and chemicals was a high priority in nearly all of the countries the WTEC panel visited. China, in particular, has a major emphasis on energy applications, especially those involving the conversion of coal to liquid fuels. Significant activities in photocatalysis, hydrogen generation, and fuel cells are being carried out in many locations. There is a general recognition that energy carriers and chemicals should be produced, and ultimately used, with as little impact on the environment as possible; catalytic solutions are thus being pursued in this framework of environmental sustainability. Catalytic production of ultra-low sulfur fuels, use of renewable carbon sources and sunlight, conversion of the greenhouse gas CO_2 to useful products, highly selective oxidation of hydrocarbons, and catalytic aftertreatment of waste streams are all being pursued vigorously around the globe. A growing area of interest is the catalytic transformation of various plant sources to energy-relevant compounds such as bio-oil (a carbonaceous liquid that can be blended into a refinery stream), biodiesel fuel, hydrogen, alcohols, etc. However, the targeted plant feedstocks depend on the native vegetation within a particular country.

Conclusions

Catalysis by nanostructured materials is an active area of research around the globe, and its rate of growth appears to be increasing faster than that of all science, presumably because of significant concerns regarding future energy security and environmental sustainability. Western Europe appears to hold the dominant position in the world in terms of research paper output, but the rapid growth of research in Asia could challenge that position in the near future. The overall investment levels in catalysis research in Western Europe and Asia appear to be significantly greater than that in the United States. Because 1996–2005 US publications were the highest cited in the world, research funds in the United States are apparently distributed effectively to the highest-quality laboratories. The overall impact of US research, however, is dampened by a much smaller output relative to Western Europe and an overall growing output from Asia. The technical themes involving catalyst synthesis, characterization, theories, and applications, have specific components that are similar to those in US research programs, but the level of research activity in particular areas often depend on regional needs.

Reference

National Science Foundation (2003). *NSF Workshop Report on Future Directions in Catalysis: Structures that Function on the Nanoscale* (NSF, Arlington), cheme.caltech.edu/nsfcatworkshop/NSF%20FinalRept%202004.pdf.

1

Overview of Catalysis by Nanostructured Materials

Robert J. Davis

1.1 Introduction to the Study

The science and technology of catalysis have played a critical role in improving our standard of living over the last century. Materials produced by catalytic technology are responsible for modern medicines, new fibers for clothing and construction, a wide variety of consumer products, cleaner-burning fuels, and environmental protection. The catalytic process used to produce high octane aviation fuel helped secure the victory of the Royal Air Force in the Battle of Britain during World War II. More recently, the widespread use of automotive catalytic converters is responsible for cleaning auto exhaust gases and improving air quality in cities around the world. The overall economic impact of catalysis cannot be overemphasized. For example, the US chemical industry, which relies heavily on catalysis, had record exports in 2007 and its first trade surplus since 2001 (Shelley, 2008). Furthermore, catalyst technologies account for over US$1 trillion of revenues in the US economy and about a third of the material GDP (Davis and Tilley, 2003).

Fundamental discoveries in catalysis have been recognized many times over the previous 100 years. The Nobel Prize (see http://nobelprize.org/) was awarded to Sabatier (1912) and Langmuir (1932) for their pioneering research on catalytic hydrogenation and oxidation reactions on metals, respectively, and Ziegler and Natta received the Nobel Prize in 1963 for their novel catalytic process to make plastics and fibers. Altman and Cech were recognized with the same honor in 1989 for their discovery of catalytic properties of RNA, while Knowles, Noyori, and Sharpless shared the Nobel Prize in 2001 for their research on chiral catalysis. Most recently, the 2007 Nobel Prize was awarded to Gerhard Ertl for his elucidation of catalytic processes on solid surfaces.

The science and technology of catalysis is more important today than at any other time in our history. Since extraction of the earth's fossil fuel resources will reach peak production in the coming century, the cost of energy and chemicals from hydrocarbon resources will continue to increase in the near future. Also, the general consensus that global climate change can be influenced by human activities has motivated researchers to find new manufacturing processes that are environmentally sustainable. Catalysis will certainly play a central role in the development of new energy sources and environmentally benign chemical technologies.

A catalyst by definition is a substance that facilitates the transformation of reactants to products through a repeated cycle of elementary steps in which the last step regenerates the catalyst to its original form (NRC, 1992). If a catalyst is present in the same phase as the reactants and products, it is called homogeneous, whereas if the catalyst is in a different phase, it is called heterogeneous. One example of a heterogeneous catalyst is found in the automotive catalytic converter and is composed of nanometer-size transition metal particles supported on a solid carrier. In that case, combustion gases first adsorb on the transition metal particles and then react to form relatively harmless products that desorb into the exhaust stream. Many industrially relevant catalysts are heterogeneous, or solid-phase, which allows for very large scale continuous operation at elevated temperatures and pressures.

Since heterogeneous catalysis is a molecular event occurring at a solid-fluid interface, the nanostructure surrounding the reactive interface, known as the active site, can significantly influence the observed rate of reaction

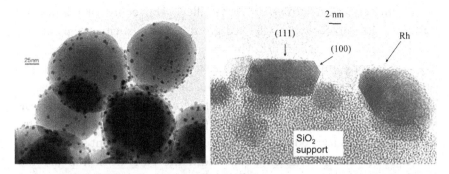

Fig. 1.1. Rhodium metal particles supported on silica carrier (*left*). The high-resolution electron micrograph (*right*) shows how small supported Rh crystallites expose low index faces (left photo courtesy of Abhaya Datye; right photo from Datye (2000), © Kluwer Academic).

(referred to as catalytic activity) and the distribution of observed products (known as selectivity). Electron micrographs of a supported transition metal catalyst are shown in Figure 1.1. The nanometer-size particles of rhodium metal are the active catalytic component, and the silica provides a high-surface-area medium to disperse the metal particles. The nature of the exposed rhodium crystal planes as well as the presence of corners, edges, and defects on the metal particles can influence the activity and selectivity of a particular reaction on the catalyst. In addition to the exposed metal atoms, the nanoscopic environment around heterogeneous catalysts can be affected by many other parameters, such as the bulk crystal structure, composition of the support, addition of a promoter, effect of micropores, polarity of the reaction medium, chirality of ligands, and nearby presence of another "active site". In essence, catalysis by nanostructured materials is a multidimensional phenomenon. Because catalysts orchestrate chemical reactions in a highly controlled spatial and temporal manner, they are considered the engines that power the world at the nanometer length scale (Davis and Tilley, 2003). Indeed, catalysts are undoubtedly "the most successful current application of nanotechnology" (Davis and Tilley, 2003).

On June 19th and 20th, 2003, a US National Science Foundation (NSF) workshop, "Future Directions in Catalysis: Structures that Function at the Nanoscale", was held at NSF headquarters in Arlington, VA (Davis and Tilley, 2003). The organizers, Professor Mark Davis (California Institute of Technology) and Professor Don Tilley (University of California, Berkeley), assembled a distinguished group of 34 participants, primarily from US academic institutions, government agencies, national laboratories, and major companies, to assess the state-of-the-art in the field and to provide visionary statements on the future directions of catalysis research. The workshop was organized around three working groups focused on (1) synthesis, (2) characterization, and (3) theoretical modeling of catalysts; these topics formed the basic framework of the current study.

Although the recommendations from the NSF workshop are too lengthy to reproduce here, several key concepts influenced the organization of the current international assessment. First, synthesis of new nanoscopic catalysts will require fundamental understanding of molecular-scale self-assembly and complex, multicomponent, metastable systems. The new tools of nanotechnology will likely play a key role in the synthesis of new catalytic structures. A second finding of the workshop is that new *in situ* characterization methods that extend the limits of temperature, pressure, and spatial resolution are needed to probe nanostructured catalysts in their working state. Finally, significant advances in theoretical descriptions of complex reactions and in models that span multiple time scales are required to improve the predictive capabilities of computation, especially in liquid phase systems. A grand challenge that emerged from the workshop is "to control the composition and structure of catalytic materials over length scales from 1 nanometer to 1 micron to provide catalytic materials that accurately and efficiently control reaction pathways" (Davis and Tilley, 2003).

Following the 2003 NSF workshop, four US government sponsors asked the World Technology Evaluation Center (WTEC) to organize a study to assess the worldwide state of the art and research trends in catalysis by nanostructured materials. The National Science Foundation, the Department of Energy (DOE), the Air Force Office of Scientific Research (AFOSR), and the Defense Threat Reduction Agency (DTRA) intended the study to:

- Identify high-impact research areas worth exploring in US R&D programs
- Clarify research opportunities and needs for promoting progress in the field

- Identify opportunities and mechanisms for international collaboration
- Evaluate the position of foreign research programs relative to those in the United States.

The sponsors agreed to maintain the same general framework as that of the 2003 NSF workshop, with an added emphasis on applications. Thus, there were four topical focus areas of the WTEC study:

(1) Design and control of synthetic catalytic structures
(2) Nanoscale characterization of catalysts in their working state
(3) Theory and simulation
(4) Applications

1.2 Approach and Methodologies

WTEC recruited a panel of eight US experts in the field, chaired by Robert Davis of the University of Virginia, to perform the assessment. Table 1.1 provides a list of the panelists and their areas of focus for the study. The table also shows others who helped arrange, conduct, and evaluate the site visits. Biographies of the panelists are given in Appendix A.

Table 1.1. Key members of the WTEC team and their roles in the catalysis study.

Name	Organization	Assignment	Technical Focus
Robert Davis	University of Virginia	Panel Chair	General Trends
Vadim Guliants	University of Cincinnati	Panelist	Applications
George Huber	University of Massachusetts	Panelist	Applications
Raul Lobo	University of Delaware	Panelist	Synthesis
Matthew Neurock	University of Virginia	Panelist	Theory & Simulation
Jeffrey Miller	BP Corporation	Panelist	*In situ* Spectroscopy
Renu Sharma	Arizona State University	Panelist	Electron Microscopy
Levi Thompson	University of Michigan	Panelist	Applications
John Regalbuto	NSF/Engineering	Sponsor/Observer	
Mike DeHaemer	WTEC	Support Staff	
Hassan Ali	WTEC	Support Staff	

A worldwide assessment of this kind normally begins with a benchmarking exercise to establish the state of the art in US R&D programs. In this case, the report from the 2003 NSF workshop on Future Directions in Catalysis: Structures that Function at the Nanoscale served as a reference point for US activities in the focus area of our study. Therefore, the WTEC study was performed in the following four phases:

(1) **Conduct a bibliometric analysis to establish world trends in publishing in the field.** The analysis examined publication trends by various countries over the last decade. Results were used to confirm and identify new sites selected by the panelists to visit. A complete summary of the bibliometric study can be found in Appendix B.

(2) **Visit a number of the world's leading university, government, and industrial laboratories.** The WTEC panelists visited 20 sites in China, Japan, and the Republic of Korea during one week in June of 2007 and 22 sites in Europe during one week in September of 2007. A complete list of those sites is provided in Table 1.2. Obviously, the panelists could not visit all of the major laboratories performing catalysis research in a two-week time frame. In a few cases, panelists met with representatives away from their laboratories, when travel logistics prevented an actual site visit. Site reports from the Asia and Europe visits are provided in Appendices C and D, respectively.

(3) **Report findings in a public workshop.** The panelists reported their findings to the US sponsors, the catalysis community, and the general public at a workshop held at the FDIC L. William Seidman Center in Arlington, VA, on November 29, 2007. The workshop, entitled "Assessment of International R&D in Catalysis by Nanostructured Materials", was a public forum in which panelists presented overall findings, specific examples from the site visits, and general conclusions. The workshop also allowed for discussion and critical review of the findings.

(4) **Compile the findings into a written report.** This document represents the written record of the study that is available to the sponsors, funding agencies, policymakers, the catalysis community, and the general public. Each panelist authored several site reports and a chapter in his or her focus area. The site reports and individual chapters were

Table 1.2. Sites visited by the WTEC the catalysis panal.

Asia

National Institute of Advanced Industrial Science and Technology (AIST; Japan)
Dalian Institute of Chemical Physics, State Key Lab. of Catalysis (PR China)
Dalian Univ. of Technology, Key Labs. of Fine Chemical Engineering, Petrochemical Technology (PR China)
Hitachi High-Technologies Corp., Naka Application Center (Japan)
Hokkaido University Catalysis Research Center (Japan)
Institute of Coal Chemistry, Chinese Academy of Sciences (PR China)
Jilin Univ., State Key Lab. for Inorganic Synth. & Preparative Chemistry (PR China)
Korea Advanced Institute of Science and Technology (KAIST; R Korea)
Peking Univ., Colleges of Chem. and Chem. and Molecular Engineering (PR China)
Photon Factory, High Energy Accelerator Research Organization (PF, KEK; Japan)
Pohang University of Science and Technology (POSTECH; R Korea)
Research Institute of Petroleum Processing (RIPP; PR China)
Seoul National University School of Chemical and Biological Engineering (R Korea)
Tianjin University, State Key Laboratory of Applied Catalysis (PR China)
Tokyo Institute of Technology, Division of Catalytic Chemistry (Japan)
Tokyo Metropolitan University Department of Applied Chemistry (Japan)
Toyota Motor Corporation, Higashi-Fuji Technical Center (Japan)
Tsinghua Univ., Inst. of Phys. Chem., Chem. Engineering Dept., others (PR China)
The Univ. of Tokyo, Depts. of Chemistry and Chemical Systems Engineering (Japan)
Univ. of Tsukuba, School of Pure and Applied Sciences, Inst. of Matls. Sci. (Japan)

Europe

Cardiff University, Heterogeneous Catalysis and Surface Science Group (UK)
Denmark Tech. Univ., Center for Individual Nanoparticle Functionality (Denmark)
Eindhoven University of Technology (The Netherlands)
ETH (Swiss Fed. Inst. of Tech., Zurich) Baiker Catalysis and Reaction Eng. Grp.
Fritz Haber Inst., Depts. of Molecular Physics and Chemical Physics (Germany)
Haldor Topsøe A/S, Research and Development Division (Denmark)
IFP/Institut Français du Pétrole (France)
Institute of Catalysis and Petrochemistry (Spain)
Inst. Charles Gerhardt Montpellier, Adv. Materials for Catalysis & Healthcare (France)
Institute of Chemical Technology (ITQ; Spain)
Institute of Research on Catalysis and the Environment of Lyon (IRCELYON; France)
Politecnico di Milano and Università degli Studi di Milano (Univ. of Turin; Italy)
Max-Plank-Institut für Kohlenforschung (Germany)
Shell Global Solutions International BV (The Netherlands)
Technical University of Munich (Germany)
University of Cambridge, Magnetic Resonance Research Center (UK)

submitted to the appropriate site hosts for review and possible correction of factual statements, as necessary. This report is available on the Web at http://www.wtec.org/catalysis/.

1.3 Report Structure

The report is generally organized along the topical areas outlined by the original 2003 NSF workshop on Future Directions in Catalysis: Structures that Function at the Nanoscale. This first chapter discusses the publication outputs of the various countries compared to the United States, compares some of the funding strategies for catalysis research in various countries, and highlights the major findings of the study. Chapter 2 by Raul Lobo discusses the important role of materials synthesis in current and future research on catalysis by nanostructured materials. Chapters 3 and 4, authored by Jeff Miller and Renu Sharma, highlight the recent advances in characterization of catalysts: Miller emphasizes the critical role of *in situ* spectroscopy in catalysis research; Sharma focuses on the use of new microscopic techniques for direct imaging of catalysts at the nanoscale. Chapter 5 by Matt Neurock discusses the growing use of *ab initio* molecular modeling and simulation in catalysis R&D and how future developments in computational methods can facilitate discovery. Chapters 6 and 7, authored by Levi Thompson and Vadim Guliants, focus on the specific applications of catalysts in production of energy carriers and chemicals from fossil fuel resources; those chapters also discuss catalytic applications aimed at pollution prevention or remediation in the utilization of fossil fuels. The final chapter by George Huber also examines how catalysts are used for energy and chemicals production, but with a particular focus on renewable feedstocks. There are three appendices:

(A) Biographies of Panelists
(B) Bibliometric Analysis
(C) Glossary of Terms and Acronyms

1.3.1 *Bibliometric analysis*

WTEC contracted with Grant Lewison of Evaluametrics, Ltd., United Kingdom, to perform the bibliometric analysis; a complete summary of the findings can be found in Appendix B. Lewison met with panel chair Davis at the University of Virginia in April of 2007 to develop a filter capable of retrieving papers appearing in the Science Citation Index (SCI) relevant to the topic of interest. The names of specialty journals (such as *Journal of Catalysis and Applied Catalysis*) and recurring title words were used to create the filter. One problem encountered with this approach was the improper retrieval of papers involving homogeneous catalysis or biocatalysis. Thus, specific elements and reactions commonly involved with solid catalysts were added to the filter. Moreover, titles with certain key words or acronyms such as DNA, protein, protease, RNA, and solar cell were eliminated from the search process. Nevertheless, the calibration of our final filter showed a precision of 0.62 and a recall of 0.78, and the results of this calibration were used to scale the output of the catalysis papers by a factor of 0.795 (= 0.62/0.78). Please see Appendix B for a detailed description of the filtering process.

Several important quantitative outputs were determined by the bibliometric analysis. First, the catalysis papers published by a particular country were counted for the 10-year period from 1996 to 2005. Compared to the world output of all science papers indexed in the SCI over the same period, catalysis papers represented about 1/60th of the world science output. Perhaps more importantly, the growth rate in catalysis papers (5.4%) exceeded that of all science papers (2.9%), which suggests a growing importance of catalysis research in the world.

Examining trends by geographic region offers additional insights into global catalysis activities. The output of catalysis papers by 13 Western European countries (Austria, Belgium, Denmark, Finland, France, Germany, Italy, Netherlands, Norway, Spain, Sweden, Switzerland, and United Kingdom) exceeded the total produced by the United States by almost a factor of two. This result is quite significant, since Lewison indicated that the output of the United States and Western Europe is comparable in most other fields of science. Another result of the analysis is the significant growth of catalysis papers from the Peoples Republic of China from 1996–2005. In 1996, the catalysis papers from the United States outnumbered those from China by a factor of 6; however, the large disparity between China and the United States was largely eliminated by 2005. A reasonable extrapolation of the exponential growth of Chinese publication output suggests that the United States may now lag China in the number of catalysis papers. Japan, which is currently the second major producer of catalysis papers in East Asia, had a much more modest rate of increase of catalysis publications compared to China, while the publication rate from the United States was essentially flat throughout the decade under analysis.

A second quantitative aspect of the bibliometric analysis involved a country's Relative Commitment (RC) to catalysis. By definition, the RC is the integer count of catalysis papers divided by the total number of science papers published by that country. Thus, the RC measures the importance of catalysis research within a particular country, with the RC value for the entire world being unity. The very high relative commitment of China (RC > 2) to catalysis research over the last decade is consistent with its exponential growth in published papers. This contrasts a very low commitment of the US to catalysis research (RC ~0.6). The relative commitment of Western Europe to catalysis varies by country, but is generally near the world average (RC = 1).

A third quantitative parameter evaluated in the analysis is the Potential Citation Impact (PCI). This parameter was calculated from the citation records of journals publishing the papers retrieved by the filter, assuming catalysis papers in the journals will be cited at the same frequency as other papers. In this particular study, the PCI was based on the expected citations of a paper in the year it was published and the four subsequent years. The papers published by US researchers appear in journals with the highest potential impact (PCI ~13), whereas the papers published in East Asia appear in journals with below-average impact factors (China PCI ~6.5, Japan PCI ~8.0, Republic of Korea PCI ~6.7). The PCI of Western European countries is near the world average of about 9 for the decade of analysis. Although the PCI associated with the US publications exceeds both East Asia and Western Europe, the high research volume of the European countries more than makes up for the difference in PCI. For example, an overall or total citation impact of a country can be calculated as the product of the PCI and the total number of papers published by a country or region. Figure 1.2 compares the total number of expected citations from papers originating from North America, Western Europe, and East Asia. The figure is consistent with the rapid growth in the number of

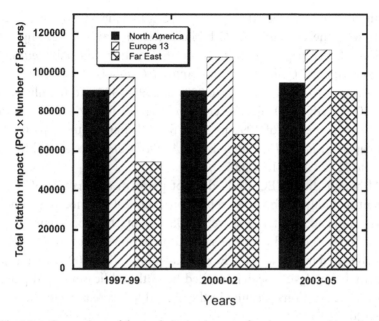

Fig. 1.2. Comparison of the total citation impact of various geographic regions.

publications from East Asia compared to the rest of the world. Although East Asian researchers tend to publish in journals with lower impact factors than US authors, the recent high volume of papers from that region resulted in a total citation impact nearly equivalent to that of the United States. Moreover, the total impact of papers from Western Europe dominated the world over the decade under study, and the gap between Western Europe and the United States widened over this time period.

The National Research Council (NRC) of the National Academies completed a benchmarking study in 2007 of the competitiveness of US research in Chemical Engineering (NRC, 2007a) and Chemistry (NRC, 2007b). Although heterogeneous catalysis is a subfield of physical chemistry, a significant fraction of catalysis research in the United States is performed in chemical engineering departments. Therefore, the benchmarking studies by the National Research Council explicitly reported on the US position in catalysis with respect to chemical engineering departments, whereas the discussion of catalysis within chemistry departments was indirect through an overall study of physical chemistry. The general findings of the NRC report on chemical engineering are completely consistent with the current WTEC bibliometric analysis. For example, the fraction of US papers published in the leading journals *Journal of Catalysis*, *Applied Catalysis A*, and *Applied Catalysis B* was 33% in the period 1990–1994, but declined to 23% in 1995–1999 and to only 15% in 2000–2006 (NRC, 2007a). The citation statistics follow a similar trend. Over the time period of 1990 to 2006, the US share of the top 50 most-cited papers in *Journal of Catalysis and Applied Catalysis* A&B fell from 27 (54%) to 12 (24%) (NRC, 2007a).

Although the WTEC study did not perform an analysis of patents relevant to catalysis, the NRC report on chemical engineering competitiveness indicates that the share of US patents that originated in the United States declined over the same time period relative to US patents that originated in Asia and the European Union (NRC, 2007a). Thus, productivity in catalysis research, gauged by both publications and patents, has shown a fractional decline in the United States relative to the rest of the world.

The results of the WTEC bibliometric analysis suggest that the United States had a significant, but not dominant, position in catalysis research over the last decade. Western Europe led the world in both total number of papers published and the total citation impact. However, the very high relative commitment in East Asia to catalysis research, particularly in China, is illustrated by the exponential growth in the number of published papers in the last decade and will soon challenge the position of both the United States and Western Europe.

1.4 Investment Models and Trends

The financial models used to support catalysis research vary widely around the world. The following sections provide examples of how catalysis is supported at some of the sites visited by the WTEC panel.

1.4.1 *The United States*

As a baseline for comparison, the US National Science Foundation and Department of Energy combined support catalysis research at universities at a level of approximately US$30 million per year, with nearly half of that originating from the Basic Energy Sciences program at DOE. The Catalysis and Biocatalysis Program at NSF, together with NSF contributions to special programs such as its Nanoscale Interdisciplinary Research Team (NIRT), Nanoscale Exploratory Research (NER), and Partnerships for International Research and Education (PIRE) programs, add approximately US$7 million per year to US university research in catalysis. In many institutions outside the United States, the clear distinction between a national laboratory and a university laboratory, as seen in the United States, is fuzzy. A fair comparison of research support must therefore include the US funding at national laboratories. The total support from the DOE Office of Science, the DOE Technology Offices, other governmental agencies, and companies for catalysis research at the US national laboratories is estimated to be about US$45 million. Although there are some aspects of catalysis research funded by the military agencies in areas such as fuel cell power and destruction of hazardous agents, the contribution of these agencies to the public research enterprise was not evaluated in this study. Moreover, energy companies, chemical companies, and catalyst manufacturers also sponsor catalysis research at some US universities. However, the major source of fundamental research support for catalysis in the United States is provided by the Department of Energy and the National Science Foundation.

1.4.2 Asia

Research in East Asia is supported by very different financial models, depending on the country. For example, the cost of a graduate student to a research grant is negligible in the People's Republic of China, since most of the educational costs are covered by other sources. Therefore, the size of research groups can be quite substantial in China. Interestingly, the very select schools in China such as Tsinghua University and Peking University limit student enrollment at all levels, which translates into modest research group sizes at these institutions. Other institutions without strict enrollment caps enable groups to reach as high as 30 students or more. The Dalian Institute of Chemical Physics (DICP) presents another model for catalysis research in China. This multidisciplinary institute of the Chinese

Academy of Sciences is home to the State Key Laboratories of Catalysis and Molecular Reaction Dynamics as well as several national projects and spin-off companies. Approximately 500 staff members are involved in catalysis research, along with a substantial fraction of their 800 graduate students. The support for the catalysis operations at the DICP is estimated to be US$15 million per year, which is nearly the same as the DOE Basic Energy Sciences catalysis budget or double the NSF catalysis budget for university research across the United States. At nearly every institution visited in China, panelists observed new research equipment being installed in the laboratories.

The financial support for catalysis in the Republic of Korea is mainly distributed through several government programs. For example, graduate students are supported in targeted departments with funds designated for "Brain Korea 21st Century" (BK21). The BK21 program is used to promote academic disciplines of national importance and provides funds for graduate education to top departments in those fields. The chemical engineering departments at Seoul National University and Pohang University of Science and Technology, where most of the catalysis research is performed at those schools, participate in the BK21 programs and receive support for most of the costs associated with graduate students. Another government program funding catalysis research in Korea is the National Creative Research Centers, which forms 5 new research centers every year after a country-wide competition in all fields of science. Funding for these centers is provided for up to 9 years. One such center, the Korea Advanced Institute for Science and Technology (KAIST) Center for Functional Nanomaterials, is supported at ~US$700,000 per year. Finally, research grants from the Korean Science and Engineering Foundation are a significant source of support for university catalysis researchers.

Funding for catalysis research at Japanese universities is provided mostly by various Japanese government agencies. For example, the University of Tokyo receives only 20–30% of its external funding from the Japanese chemical industry, with the rest presumably originating from government sources. The research funding per full professor at the University of Tokyo is about US$600–800 thousand, not including summer faculty salary, student stipends, tuition, and overhead. Since the Japanese government provides funding directly to graduate students for stipends and tuition, the external research funds raised by faculty can be used for a variety of needs in the laboratories. The Japanese Ministry of Education, Culture, Sports, Science and Technology recently established a fund to create "The 21st Century Centers of Excellence" program, which covers all academic fields and gives priority support for global research and education centers. Almost a quarter of University of Tokyo professors participate in this program. In addition to the catalysis research at universities, Japan also supports the activities at national laboratories such as the National Institute of Advanced Industrial Science and Technology (AIST), which is the largest public research organization in Japan. Catalysis plays a central role in several of the research themes within the institute, including centers in Biomass Technology, Hydrogen Industrial Use and Storage, and New Fuels and Vehicles Technology, as well as institutes on Innovation in Sustainable Chemistry and Energy Technology. The Institute for Innovation in Sustainable Chemistry had a total budget in FY2006 of US$18.2 million for a technical staff of 86 scientists, 35 postdoctoral researchers, and 60 technicians.

1.4.3 Western Europe

Support for catalysis research in the European Union varies widely by country and by institution. The Fritz Haber Institute (FHI) in Germany is funded primarily (~60%) by the Max Planck Society, with grants from the German Science Foundation, the European Union, and other sources making up the rest. Only about 5% of the Fritz Haber Institute's catalysis research budget is derived from industrial sources, with most of those funds being directed to the catalysis research in the Department of Inorganic Chemistry. The overall annual budget for personnel and operating costs at the FHI is US$35–40 million, which supports 5 director positions, 148 technical staff positions, 80–100 PhD students (many with outside funding), and 35 support staff. Although the FHI pursues fundamental research in a variety of fields, the institute has a current focus on catalysis, chemical and physical properties of interfaces, molecules, clusters, and nanostructures. In contrast to the FHI, the catalysis activities at the Technical University of Munich are funded by competitive government grants, noncompetitive government support, and industrial contracts, shared almost equally among the sources.

Catalysis research in Spain is focused at the Institute of Catalysis and Petrochemistry (ICP) in Madrid and the Institute of Chemical Technology (ITQ) in Valencia. Both institutes are funded by the Spanish Council for Scientific Research (CSIC) at substantial levels. The ICP is the largest catalysis research institution in Spain and consists of 4 departments: Structure and Reactivity, Applied Catalysis, Catalytic Process Engineering, and Biocatalysis. To provide a sense of scale, the Structure and Reactivity department has an annual budget of about US$2 million, half of which is provided by the CSIC. The department supports 10 staff scientists, 4 postdoctoral researchers, 17 PhD students, 4 contracted graduate students, and an administrative person. The ITQ has a staff of about 100 people (64 researchers and technical staff members, 28 doctoral students, and 21 postdoctoral researchers) supported by an annual budget of about US$6 million. The unique financial structure of the ITQ has enabled fantastic expansion from its inception in 1991 and accounts for about 600% growth in income over the last decade. Almost 60% of that income is derived from contracts with companies and licensing fees generated by patents developed at ITQ.

The French government has established the following 7 priority areas of research for its country: (1) Biology and Health; (2) Ecosystems; (3) Energy and the Environment (including CO_2 minimization, hydrogen and fuel cells, bioenergy, transport, and clean vehicles); (4) Engineering Processes (including chemistry and processes, materials and processes); (5) Physics and Information; (6) Humanities and Society; and (7) Basic Science (Chemistry and Multidisciplinary Studies). Although catalysis has a major role in 3 of the 7 priority areas and is clearly a field with substantial support in France, there was an indication that research funding may decline in the future. Nearly a third of academic catalysis research in France occurs at the Institute of Research on Catalysts and the Environment of Lyon (IRCELYON). The annual budget for IRCELYON is ~US$16 million, with over half of that budget dedicated to 113 permanent positions funded by the Centre National de la Recherche Scientifique (CNRS), government, and the University of Lyon. The rest of the budget is for equipment, supplies, and support for 68 PhD students and 23 postdoctoral researchers. A completely different model for catalysis research is practiced by the Institut Français du Pétrole (IFP). The IFP performs both technology development (75%) and basic research (25%), and serves as an interface between academic and industrial groups. It is supported by the French government (two-thirds) and licensing revenues (one-third) from inventions developed at the institute. About US$60 million of the annual budget is dedicated to catalysis research in support of the activities of 200 staff members and about 60 PhD students.

The Netherlands and Denmark have made substantial recent investments in catalysis research. For example, The Netherlands Organization for Scientific Research selected the National Research School Combination-Catalysis (NSRC-Catalysis) in 1998 as one of six special programs to be funded in a competition with all fields of science. This program supported catalysis research activities at 8 different universities over the last decade and has been renewed at multimillion-dollar-per-year levels through 2013. Last year, the program supported the activities of 50 researchers, with doctoral and postdoctoral researchers comprising the majority of the personnel. Very recently, the Netherlands Institute for Catalysis Research (a virtual institute for catalysis research and education) was awarded a center grant in the area of bio-renewables conversion for more than US$40 million over the next eight years. The center, called CATCHBIO, couples universities across the Netherlands with a variety of industrial collaborators. The catalysis activities in Denmark are focused at the Technical University of Denmark (DTU) and collaborations with close-by Haldor Topsøe. The research is directed by several well-funded centers at DTU, including the Center for Atomic-Scale Materials Design (CAMD) and the Center for Individual Nanoparticle Functionality (CINF). The funding for CAMD, which is generally recognized to be one of the world's leading centers in computational catalysis, is reported to be US$14 million over the 2006–2011 time period, excluding computer grants and funds for permanent staff of 7 faculty members and 6 support persons; over one-third of the support is provided by the Lundbeck Foundation. The CAMD has 8 faculty members, 12 postdoctoral researchers, 17 PhD students, and 7 support persons.

Italy has about 500–600 catalysis researchers in catalysis that participate in a variety of national and European consortia. The Interuniversity Consortium for the Science and Technology of Materials is comprised of 44 universities performing research in materials science in connection with chemistry, engineering, and nanotechnology; one of the eight section topics of this consortium is relevant to catalysis and reactive interfaces. The estimated budget for the catalysis activities is about US$2 million per year. Also, Italy participates in the European IDECAT initiative (Integrated Design of Catalytic Nanomaterials for a Sustainable Production), which was formed in 2005 and is funded at approximately US$14 million over a five-year period. IDECAT involves 37 laboratories from 17 institutions in 12 countries with about 500 participants overall and has goals to integrate top-level EU expertise in catalysis and to create a critical mass of researchers that will enable a step-change in catalysis R&D in Europe.

The United Kingdom has about 6 groups substantially involved with catalysis and surface science; the WTEC team visited sites at Cardiff University and Cambridge University. The 65 people at Cardiff University (7 academics, 13 postdoctoral researchers, and 45 students) are funded by more than US$10 million in external grants, which are derived primarily from government sources (60%) and the rest coming from industrial contracts. The Surface Science activities at Cambridge University are supported by over a US$1 million per year by the government, with additional industrial grants, to support a total staff of 1 senior professor, 1 research fellow, 7 postdocs, and 11 PhD students.

1.4.4 Summary

The different financial models used throughout the world are effective at supporting catalysis research; no single system can be declared as the best. However, certain aspects of each deserve close examination for possible adoption in the United States. For example, the sustained level of baseline funding provided to a science-oriented institution such as the Fritz Haber Institute allows for effective utilization of the highly trained research staff that is necessary to build the world's next-generation scientific instruments. However, a stable funding stream is not the only unique feature that allows the institution to function at a very high level scientifically. Justification of research directions at the FHI is based on a recommendation from an external evaluation committee of scientific experts from around the world instead of classical numerical outputs such as papers and patents. This stable source of funding for long-term research goals at the FHI contrasts with the typical three-year granting cycle at the US government agencies, which also require yearly justifications for grant expenditures. This difference in funding strategy between the US government and the FHI was highlighted during the WTEC visit as a major reason why certain kinds of research and instrument development are unique to the FHI.

The consortia model developed within and throughout the European countries is another strategy for funding research. These consortia tend to be most effective within a single country, presumably because of the ease of communication, shared cultural identity, and geographic proximity of participating institutions. Although pan-European initiatives have also experienced success, a significant fraction of funding in those initiatives is dedicated to travel and collaboration instead of directly supporting research.

The European and Asian countries have done an excellent job combining academic research with national laboratory activities. The combination is almost seamless at many institutions, with principal investigators, professors, doctoral students, and research staff members working together on common goals. There appears to be a more significant separation between academic catalysis research in the United States and its complement at the mission-oriented DOE national laboratories.

The support of doctoral students and postdoctoral researchers on US research grants tends to consume the major fraction of direct charges. In countries like China and Korea, government programs that fund students directly for their graduate education in areas of national importance remove the need to fund students directly on research grants. Although it is unclear whether or not that system is superior to funding students directly on grants, it does appear to be similar to the graduate training grant programs in the United States that are common in other fields.

The WTEC panelists found significant industrial collaborations at most of the sites and had the impression that industrial support is more prevalent outside of the United States. The reasons appear to be many, such as a lower cost of performing research outside of the United States, a more cooperative intellectual property environment outside of the United States, and possible access to emerging world markets for the next users of catalytic technologies. Moreover, the panel noted the high quality of pilot plant facilities and catalytic reactor systems capable of industrial operating conditions located within academic settings. The infrastructure for fundamental research on catalyst synthesis and characterization combined with industrially relevant catalyst testing appears to exceed that of typical academic catalysis laboratories in the United States.

Routine equipment for catalysis research such gas chromatographs, reactors, vacuum chambers, adsorption systems, bench-top spectrometers, as well as major research instruments such as electron microscopes, nuclear magnetic resonance spectrometers, and X-ray diffractometers is well represented worldwide. In addition, researchers appear to have ready access to synchrotron light sources for advanced *in situ* characterization of catalysts. However, the panelists reported a significant concern with the aging catalysis infrastructure in the United States compared to that observed in East Asia and Europe. The difficulty US researchers currently have with acquisition of new instruments routinely used for catalysis research could severely impact US competitiveness in the very near future.

1.5 General Observations

As stated in the NRC report on chemical engineering competitiveness, catalysis is often associated with large-scale chemical, petrochemical, or oil refinery processes, which are areas of rapid growth in Asia but are fairly stagnant in the United States and the European Union (NRC, 2007a). Although Asia's rapidly expanding economy accounts for its major growth in catalysis research, the reasons for the dominant position of Western Europe in catalysis research output are not so straightforward.

Several key factors are likely to contribute to this result. First, catalysis is generally viewed outside of the United States as a fundamental science that enables discovery and development of technology in a variety of energy- and chemicals-related fields. Therefore, catalysis research is pursued in chemistry and physics departments throughout Europe, with much smaller levels of activity in European chemical engineering departments. Within the United States, heterogeneous catalysis has been viewed as a mature field that is studied mainly in chemical engineering departments. Chemistry departments in the United States tend to support much more fundamental studies involving model surfaces in pristine environments, homogeneous or single-site molecular catalysis, and biocatalysis. This artificial division may allow heterogeneous catalysis in the United States to be incorrectly perceived as an applied field of research instead of one with the potential for fundamental discovery.

Another reason European catalysis appears to have a dominant position is the close coupling in Europe of universities and national laboratories with industry. Companies within several of the countries the WTEC panel visited appear to campaign for catalysis research at the national level. This kind of advocacy support from companies, together with research contracts involving universities and government labs, indicates a significant role of the private sector in setting research directions. Moreover, the current intellectual property environment in Europe appears to foster university-company relations, at least more so than in the United States. However, there are indications that European universities are beginning to explore the position of many US universities on IP ownership and may soon move in the direction of the United States where universities routinely negotiate for ownership of IP.

Finally, the overall level of investment in catalysis research in Europe simply appears to be higher than that in the United States. Although the United States is starting to see large block grants from companies such as BP and the Dow Chemical Co. for catalysis research at universities, the combined investment of the US government (mainly NSF and DOE) and the industrial sector in university and national laboratory research appears to be far below that of the European countries.

The WTEC panelists observed the use of high-throughput instrumentation in both Asia and Europe. For example, robotic synthesis of zeolite materials and high-throughput reaction testing allowed for discovery of new catalytic materials with unique properties at the ITQ in Valencia, Spain. Also, the Center for Microchemical Process Systems at KAIST in Korea makes extensive use of high-throughput screening methods for discovery of new materials. New instrumentation recently purchased at institutions throughout Asia and Europe was aimed at rapid analysis of catalyst samples. Since research abroad is often carried out in major centers of activity, some of the instrumentation was actually invented and constructed in-house. This model contrasts with the operation of most academic laboratories in the United States, which have neither in-house expertise for tool creation nor the resources for tool construction.

The panel also noted the effective use of permanent research staff positions in laboratories outside of the United States. In the vast majority of sites visited, including those associated with institutes and universities, the number of dedicated staff members in support of the research activities appeared to far exceed those associated with US laboratories. The employment of highly skilled technical staff members at the Fritz Haber Institute was highlighted during the visit as a major reason why next-generation research instruments can be designed and built within the facility. The instruments were too complex to be constructed solely by graduate students and short-term postdoctoral researchers. The current funding models in the United States do not support the same level of technicians, and academic institutions do not appear to have funds available for additional positions; in fact, many US researchers are experiencing a decrease in technical staff at their universities.

1.6 Technical Themes of the Study

The subsequent chapters in this report present the findings of the panel with respect to the four technical themes of the study: synthesis, *in situ* characterization, theory and simulation, and applications. Regarding catalyst synthesis, new methods to produce highly structured solids such as zeolites and carbons having both micropores and mesopores were being actively pursued in several countries. Moreover, preparation and stabilization of metal and metal oxide nanoparticles with controlled facets, sizes, and compositions are also areas of wide interest. Characterization of nanostructured catalysts by environmental electron microscopy is revealing how nanoscopic features are affected by the surrounding environment; this advanced method of observation will be pursued more aggressively with the increasing availability of commercial instruments. *In situ* spectroscopy has now become a routine method for catalyst characterization around the world. However, rapid improvements in temporal and spatial resolution of many spectroscopic methods as well as adaptation of methods to allow for interrogation of catalysts under industrial reaction conditions are enabling an unprecedented level of new information to be collected on the structure of catalysts in their working state. Theory and modeling have gained universal acceptance as necessary tools for advancing catalysis science. Improvements in method accuracy, computational speed, and model development have moved theory to a position alongside experimentation in many laboratories. Theory has excelled in the prediction of atomic structure and spectroscopic features of catalytic materials. Although the prediction of reaction kinetics is still developing, theory is being used to suggest novel compositions to improve catalytic performance.

The key applications stimulating most of the catalysis research worldwide were related to energy and the environment. Conversion of nonpetroleum feedstocks such as coal, natural gas, and biomass to energy and chemicals was a high priority in nearly all of the countries visited. China, in particular, has a major emphasis on energy applications, especially those involving the conversion of coal to liquid fuels. Significant activities in photocatalysis, hydrogen generation, and fuel cells are carried out in many locations. There is a general recognition that energy carriers and chemicals should be produced, and ultimately used, with as little impact on the environment as possible; catalytic solutions are thus being pursued in this framework of environmental sustainability. Catalytic production of ultra-low sulfur fuels, use of renewable carbon sources and sunlight, conversion of the greenhouse gas CO_2 to useful products, highly selective oxidation of hydrocarbons, and catalytic after-treatment of waste streams are all being pursued vigorously around the globe. A growing area of interest is the catalytic transformation of various plant sources to energy-relevant compounds such as bio-oil (a carbonaceous liquid that can be blended into a refinery stream), biodiesel fuel, hydrogen, alcohols, and so forth. However, the targeted plant feedstocks depend on the native vegetation within a particular country.

1.7 Conclusions

Catalysis by nanostructured materials is an active area of research around the globe. Its rate of growth appears to be increasing faster than that of all science, presumably because of significant concerns regarding future energy security and environmental sustainability. Western Europe currently holds a dominant position in the world in terms of research paper output, but the rapid growth of research in Asia over the previous decade could challenge that position in the near future. The overall investment levels in catalysis research in Western Europe and Asia appear to be significantly greater than that of the United States. Since recent US publications are the highest-cited in the world, research funds in the United States are distributed effectively to the highest-quality laboratories. The overall impact of US research, however, is dampened by a much smaller output relative to Western Europe and the growing output from Asia. The technical themes involving catalyst synthesis, characterization, theories, and applications have specific components that are similar to those in US research programs, but the level of research activity in particular areas often depend on regional needs.

References

Datye, A. K. (2000). Modeling of heterogeneous catalysts using simple geometry supports, *Topics in Catalysis*, 13, pp. 131–138.
Davis, M. E., and Tilley, D. (2003). *Future Directions in Catalysis Research: Structures that Function on the Nanoscale* (National Science Foundation: Washington, DC). http://cheme.caltech.edu/nsfcatworkshop/NSF FinalRept 2004.pdf.
National Research Council (1992). *Catalysis Looks to the Future* (National Academies Press: Washington, DC).
National Research Council (2007a). *International Benchmarking of U.S. Chemical Engineering Research Competitiveness* (National Academies Press: Washington, DC).
National Research Council (2007b). *The Future of U.S. Chemistry Research: Benchmarks and Challenges* (National Academies Press: Washington, DC).
Shelley, S., Ed. (2008). U.S, chemical trade balance turns positive in 2007, *Chem. Eng. Prog.*, 104, p. 12.

2

Synthesis of Nanostructured Catalysts

Raul F. Lobo

2.1 Introduction

As defined in Chapter 1, "a heterogeneous catalysis is a molecular event occurring at a solid-fluid interface [where] the nanostructure surrounding the reactive interface, known as the active site, can significantly influence the observed rate of reaction (referred to as catalytic activity) and the distribution of observed products (known as selectivity)". This definition points to the need to control the molecular structure of a heterogeneous catalyst at the nanometer length scale to successfully prepare catalysts that are both active and selective. The catalysis community endeavors to accomplish this objective by the creative use of diverse synthesis methods. This chapter describes important examples of successful control of catalyst nanostructure and its impact on catalytic activity and selectivity, as observed by the WTEC panel during its visits to laboratories in Asia and Europe. The aim of this review is not to be exhaustive, but rather to illustrate the innovative ways in which scientists and engineers are successfully controlling atoms and molecules to self-organize in cooperative assemblies with meaningful catalytic functions. The selected examples of nanostructured catalysts described here are ones that have shown interesting or unique activities.

The definition of heterogeneous catalysis given above does not reveal the difficulty of the task that is the preparation of successful and novel catalysts. In particular, in addition to catalytic activity and selectivity, practical catalytic materials must be very stable at reaction conditions. Practical industrial reaction conditions often require high temperatures and strong oxidizing or reducing environments — environments that must be withstood for extended periods of time by the nanostructured catalysts. In addition to structural stability, long-term catalytic activity requires that impurities (either in the feed or by-products of the reaction) do not accumulate on the catalyst surface and block access by the reacting molecules to the active site. This chapter discusses mainly the activity and selectivity of novel nanostructured catalytic materials because these are the properties readily controlled through synthesis. The issues of stability are nevertheless important and are discussed in more detail in other chapters.

First, some context and definitions are provided here for some of the ideas discussed later by describing an example of a nanostructured industrial catalyst already used in industry, platinum (Pt) nanoparticles supported on zeolite K-L (Treacy, 1999). This example epitomizes the importance of nanoscopic length scales in catalytic materials and is one to which other materials described below can be compared. Zeolite L has a one-dimensional pore system with pore windows of ~7.5 Å and cages between the windows of ~11 Å. This material is an excellent catalyst for the reforming of n-hexane and n-heptane into benzene and toluene (McVicker *et al.*, 1993), and it is used commercially for this purpose in several oil refineries. The presence of Pt nanoparticles in the pores of this zeolite is revealed by Z-contrast TEM in the left image of Figure 2.1. This figure also illustrates the diversity of Pt clusters in the zeolite pore (right image of Figure 2.1) in the fresh catalysts. After time-on-stream, the smaller particles aggregate to make particles similar to type B or H. The right side of Figure 2.1 also illustrates that a portion of the active sites can be inaccessible to reacting molecules if two large clusters block the one-dimensional pores of the zeolite. There is then an optimal loading of metal on the catalysts that depends on the average crystal size of the zeolite

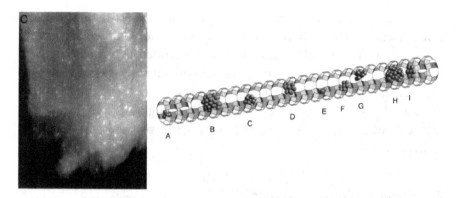

Fig. 2.1. (*Left*) A Z-contrast TEM image of zeolite Pt/K-L after reduction and reaction in oil; (*right*) illustration of the location and variable size of Pt clusters in the zeolite pores. In the optimum catalyst, each pore will contain only one particle (of the type B or H) in the pores to maximize access to the active metal surface (Treacy, 1999). As depicted, clusters B and H block access to the clusters C-G.

support. In the best-case scenario, there will be one metal cluster per pore in each of the crystal pores.

Pt/Zeolite L catalysts are prepared by impregnation, where cationic mononuclear Pt species are dissolved in water and added to the zeolite support. The nanoparticles are formed upon heating this sample in a reducing atmosphere (McVicker *et al.*, 1993). During the heating and reduction, the cationic species migrate into the zeolite pore, decompose, and are reduced. Platinum clusters are formed by migration and aggregation of Pt atoms. Where is the nanostructure "design" in this synthesis? In this case, it is the selection of the zeolite with the precise pore size and shape, and the composition needed to stabilize metal clusters in the pores. The Pt clusters do indeed self-assemble from the molecular species during the activation process. Choosing the "right" zeolite and the proper activation protocol are essential steps to make this nanostructure possible. It must be recognized, too, that in this system the thermodynamically favored state is the formation of large metal particles on the outside of the zeolite crystal. The synthesis process is crucial to capture the metastable disperse nanoparticle phase for long periods of time and to avoid the direct conversion of the precursors into large metal particles.

In the examples of catalyst synthesis that follow, self-assembly of inorganic precursors on a nanostructured support, or the use of self-assembled moieties (such as micelles and rods), are exploited in various ways to achieve materials with novel structures and with interesting catalytic properties arising from the new structure. Self-assembly of molecular entities plays a role in almost every case and can be thought of as the paradigm that connects the diverse materials and synthesis methods discussed throughout.

2.2 Rhenium Clusters in Zeolite ZSM-5

Professor Iwasawa at the University of Tokyo (see Site Reports-Asia on the International Assessment of Research in Catalysis by Nanostructured Materials website, www.wtec.org/private/catalysis; password required, contact WTEC) recently reported a new rhenium-based catalyst with exceptional selectivity for benzene hydroxylation to form phenol using molecular oxygen as the oxidant (Bal et al., 2006). The catalyst is prepared by a chemical vapor deposition method using trioxomethylrhenium as the precursor. This rhenium precursor reacts with the acid sites of the zeolite, forming isolated species of rhenium trioxide bound to the zeolite framework. The catalyst is then activated in the presence of ammonia, leading to a reorganization of the intracrystalline rhenium forming isolated Re clusters. The highest selectivity is obtained by flowing a small amount of ammonia along with benzene and oxygen, and the only by-product detected is CO_2.

Both zeolite structure and composition are very important to form the selective Re clusters. Zeolites beta, mordenite, ZSM-5, and USY were investigated, and only the catalysts prepared on zeolite ZSM-5 showed high activity and selectivity. The composition is also very important, because the selectivity of the catalysts increases as the amount of aluminum in the ZSM-5 framework increases (from 48% to 88% selectivity) (Tada et al., 2007). An extended-X-ray absorption spectroscopy investigation of the samples and their evolution with pretreatment revealed that the initially mononuclear species aggregates upon heating and — in the presence of ammonia — forms a very well defined rhenium oxynitride cluster (Figure 2.2). The rhenium assemblies

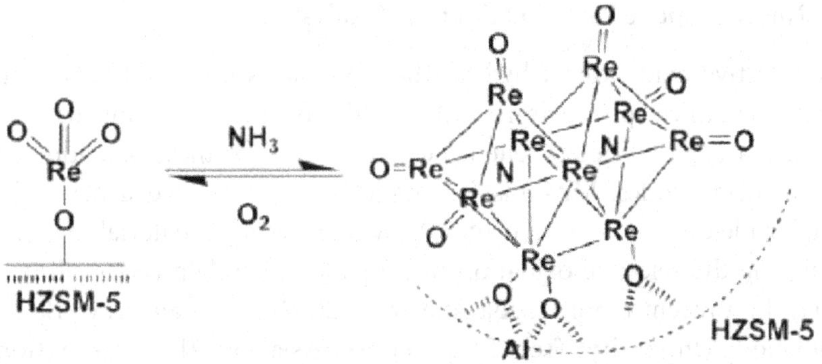

Fig. 2.2. Rhenium oxynitride clusters form in H-ZSM-5 zeolites in the presence of ammonia (Bal et al., 2006).

are edge-sharing octahedra with nitrogen atoms in the center and oxygen atoms capping the corners.

The complex has formally a positive charge and is anchored to the zeolite by electrostatic forces. This structure shows why a small ammonia pressure is required to maintain high activity and selectivity with the cluster. The ammonia provides enough nitrogen background pressure to keep the stabilizing nitrogen atoms in the cluster from leaving the active sites. If the flow of ammonia is stopped, the clusters eventually decompose and catalytic activity is lost. Again, it is important to recognize that only ZSM-5 gives acceptable catalysis levels. It is a zeolite that incidentally turned out to have the most suitable channel dimensions to allow the growth and assembly of the rhenium into a 10-unit cluster — but no bigger. Could this have been predicted from the outset? Probably not, because it was not known *a priori* what the structure was of the active site. The catalyst was prepared following a hint from previous reports suggesting rhenium inside zeolites could be a selective catalyst for benzene hydroxylation (Kusakari et al., 2004). By the systematic use of well-organized synthesis techniques (CVD, impregnation, etc.) and activation protocols, Iwasawa's group discovered this very interesting material. It was the combination of good chemical intuition and a well-organized synthesis research plan that allowed them to find this outstanding example of nanostructured catalysis.

2.3 Novel Propene Partial Oxidation Catalysts

The selective oxidation of hydrocarbons accounts for about 25% of the organic chemical products manufactured worldwide. Among the most important types of heterogeneous catalysts used for these oxidations are mixed metal oxides. Mitsubishi Chemicals has developed a new mixed metal oxide catalyst of composition MoVTe(Sb)NbO, a material capable of catalyzing the selective oxidation of propane to acrolein (Grasselli *et al.*, 2003). In a recent report (Sadakane *et al.*, 2007), Ueda and coworkers at Hokkaido University (see Site Reports-Asia on the International Assessment of Research in Catalysis by Nanostructured Materials website, www.wtec.org/private/catalysis; password required, contact WTEC) describe the synthesis of a new material with many structural similarities to the Mitsubishi catalyst. The basic structure of the Mitsubishi catalyst is depicted in Figure 2.3(a). Here, 6- and 7-ring pores are bounded by $\{MO_6\}$ octahedral and pentagonal $\{(M)M_5O_{27}\}$ units that are comprised of a heptagonal $\{MO_7\}$ unit surrounded by edge-sharing $\{MO_6\}$ octahedra (DeSanto *et al.*, 2004). Each unit cell contains four 7-ring pores and four 6-ring pores.

The new structure discovered by Ueda's group is comprised of the same building units arranged with a different connectivity. The result (Figure 2.3(b)) is a material also containing 6- and 7-ring pores. The new

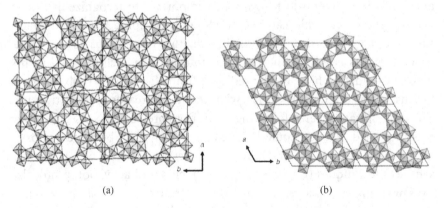

Fig. 2.3. (a) Tetragonal structure of mixed metal oxide Mitsubishi catalyst, (b) newly discovered trigonal mixed metal oxide discovered by Ueda at Hokkaido University (Sadakane *et al.*, 2007).

structure has three 7-ring pores and only two 6-ring pores per unit cell, the result of a different concentration of the $\{MO_6\}$ octahedra with respect to the pentagonal units. The catalytic tests performed on both catalysts show that their activities are remarkably similar for the acrolein-to-acrylic-acid reaction (the 2nd step in the propene oxidation process). These similarities indicate that the layered structure and the presence of both 6 and 7 rings are important to achieve high activity and selectivity at low temperatures. Crude Mo_xV_y oxides were much less active for this reaction.

A Raman and UV/V investigation of the synthesis and assembly of the tetragonal and trigonal catalysts reveals important details about the formation of intermediate building units during the synthesis process. These studies show that by controlling the pH carefully during the synthesis of the materials, pentagonal units $\{Mo(Mo)_5\}$ are formed in the precursor solution. This conclusion is inferred from the Raman signatures of the solution, consistent with the formation of polyoxomolybdates that contain the pentagonal unit. UV/Vis spectra are also consistent with the presence of polyoxomolybdates containing the pentagonal unit. These studies show that synthesis of the tetragonal and trigonal catalysts depends on the formation of nanoscale building blocks that can self-assemble into an ordered solid upon the hydrothermal treatment of the synthesis solution.

2.4 Mesoporosity Designed into Microporous Catalysts

Selective heterogeneous catalysts usually operate with high selectivity only within a narrow temperature window. Below some minimum temperature, chemical reactions proceed too slowly to be of practical value. Above an effective maximum temperature, secondary reactions become kinetically dominant and make the catalyst impractical. In zeolites and other microporous catalysts, the temperature window of operation for catalytic chemistry is coupled to transport (physical) processes. That is, sometimes the reactants or products diffuse slowly within the catalyst particles, greatly reducing the effectiveness of the catalyst.

This problem is widely recognized as a practical limitation to the application of zeolites, and many attempts and strategies have been devised to overcome it. The obvious one is to reduce the size of the crystallites. This often works, but crystal size reduction also decreases the thermal stability of the zeolite and increases the ratio of external/internal surface areas, promoting unselective reactions that occur on the crystal exterior. Recently, much effort has gone into developing materials that are microporous at one level but also mesoporous at another level, such that the resistance to diffusion is drastically reduced. During visits to Asia and Europe, the WTEC panel observed various strategies to form meso-microporous materials, a promising approach to solve this problem.

2.4.1 Micro-mesoporous zeolites by design of organic-inorganic surfactants

The group of Prof. Ryoo of the Korea Advanced Institute of Science and Technology (KAIST; see Site Reports-Asia on the International Assessment of Research in Catalysis by Nanostructured Materials website, www.wtec.org/private/catalysis; password required, contact WTEC) has developed an innovative strategy for the synthesis of zeolite crystals containing within them mesoporous channels (Choi et al., 2006). The key developments are the design of a new structure-directing agent containing an alkoxysilane moiety to anchor the molecule to the zeolite, a quaternary ammonium group to provide solubility in the synthesis gel, and an alkyl chain to promote aggregation of the organic structure directors. Figure 2.4 shows an example of a prototypical molecule used by the Ryoo's group.

Using this approach, the Ryoo group has been able to prepare several zeolite materials with extraordinary crystal mesostructure. Figure 2.5 shows a typical example of the dramatic change that is obtained in crystal mesostructure. The classical ZSM-5 morphology has particles with a "coffin" shape and with very smooth surfaces (Lai et al., 2003), completely different from the morphology of Figure 2.4.

Analysis of the mesoporosity of these materials using adsorption studies clearly shows highly monodisperse pores, consistent with aggregation of the surfactant structure-directing agents into rods as the zeolite crystal grows towards its final shape. Further indication of the order of the mesoporosity is obtained from X-ray powder diffraction studies that show a relatively narrow peak at low angles (less than $1° \, 2\Theta$). This peak arises because of the correlation between center-to-center distances of contiguous mesopores.

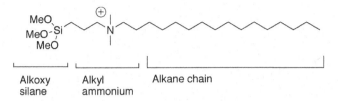

Fig. 2.4. Prototypical molecule.

Fig. 2.5. Example of the crystal morphology of zeolite ZSM-5 obtained using an organic structure-directing agent similar to the one depicted in Figure 2.4 (courtesy of Ryong Ryoo, KAIST).

Perhaps the most dramatic effect is observed on the catalytic selectivity and activity of these materials (Srivastava *et al.*, 2006). One of the reactions where these new materials have shown promise is in the synthesis of jasminaldehyde. Figure 2.6 shows the reaction of interest starting from benzaldehyde and heptanaldehyde. A mesoporous ZSM-5 zeolite shows excellent activity and selectivity (98% and 98%, respectively) towards the formation of jasminaldehyde (Choi *et al.*, 2006; Srivastava *et al.*, 2006). These numbers can be compared to traditional H-ZSM-5 zeolites

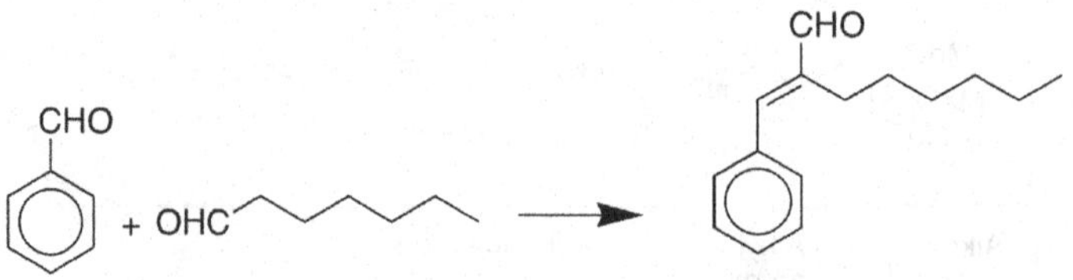

Fig. 2.6. Synthesis of jasminaldehyde using mesoporous zeolites.

(3.9% activity and 69% selectivity) and mesoporous aluminosilicates (MCM-41-type materials) that have still lower activity (25%) and lower selectivity (79%) than the mesoporous zeolites. This difference in activity indicates not only that the mesoporosity of the material helps increase catalytic activity, but also that there is some structural change on the mesoporous crystal surface (external to the micropores, but internal to the mesopores) that allows for this large improvement in catalytic selectivity. The synthetic approach developed by the KAIST group is very flexible. At this point, it seems that only a small portion of many structural variations of the molecules have been explored. This is a promising route to discovering new nanostructured catalysts with improved selectivity and activity.

2.4.2 Micro-mesoporous zeolites from carbon templates

Christensen and coworkers at the Technical University in Denmark (see Site Reports-Europe on the International Assessment of Research in Catalysis by Nanostructured Materials website, www.wtec.org/private/catalysis; password required, contact WTEC) have developed a completely different approach to making zeolite catalysts containing mesopores (Kustova *et al.*, 2004; Christensen *et al.*, 2005; 2007). The approach is called the carbon templating method, whereby zeolites are synthesized using a highly concentrated gel in the presence of carbon materials such as carbon black particles (~12 nm in diameter), carbon fibers, carbon nanotubes, etc. During crystal growth, zeolite crystals grow around the carbon structures in the synthesis gel, engulfing the carbon particles. After the zeolite synthesis has been completed, the samples are heated in the presence of oxygen and the carbon is burned completely, leaving spaces within the crystal with irregular orientations and locations but relatively uniform in

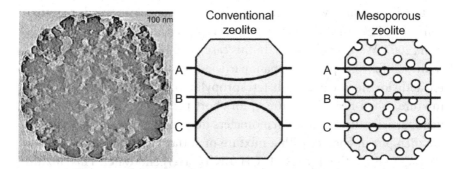

Fig. 2.7. (*Left*) TEM image of zeolite silicalite-1 prepared using carbon black particles in the synthesis gel (Janssen *et al.*, 2003); (*right*) diagram illustrating the concentration profile of benzene (A), ethylene (B), and ethylbenzene (C) on a conventional zeolite crystal and a mesoporous zeolite prepared by the carbon templating process (Christensen *et al.*, 2007).

size. The TEM image in Figure 2.7 is of a crystal of ZSM-5 after calcination and removal of the carbon. The mesopores can be observed clearly in the image. This method is very flexible and is widely applicable to the synthesis of many zeolite structures. It can yield a large fraction of mesoporous volume (up to 1.0 cc/g).

Recently, Christensen *et al.* (2007) have investigated simultaneously the catalytic activity and diffusion rates of benzene and ethylbenzene in mesoporous ZSM-5 crystals. The catalytic tests show that for zeolites of nominally the same crystal size, the reaction rate for benzene alkylation is much faster for the mesoporous zeolite than for the conventional zeolite crystal. Using the classical ideas of diffusion-controlled transport in catalyst pellets, they derive activation energies for the diffusion of reactants and products. Their analysis indicates that the effect of the mesopores is to accelerate the transport of reagents into the crystal (and products out of the crystal). Their analysis provides an explanation for increases in both selectivity and activity for the reaction investigated.

2.4.3 *Micro-mesoporous catalysts by assembly of nanoparticles*

The group of Feng-Shou Xiao at Jilin University (see Site Reports-Asia on the International Assessment of Research in Catalysis by Nanostructured

Materials website, www.wtec.org/private/catalysis; password required, contact WTEC) has devised a different approach (Li et al., 2005; Tang et al., 2007; Wang et al., 2005). In this case, the investigators first prepare a solution of zeolite precursor nanoparticles by mixing, for example, water, tetraethylorthosilicate, and tetrapropylammonium. This particular mixture, when heated to ~100°C for short periods of time, gives rise to nanoparticles that are a few nanometers in diameter. These nanoparticles are then put in contact with a mixture of surfactants: Pluronic P123 and $(C_3F_7O-C_3F_6O)_2CFCF_3CONH (CH_2)_3N^+(C_2H_5)_2$ and water. This mixture is stirred and heated under hydrothermal conditions at 180°C. The product contains a very well-defined mesoporous structure similar to the one of mesoporous silica USB-1. The composite material is called JLU-20.

Using a combination of nitrogen adsorption isotherms, nuclear magnetic resonance and infrared spectroscopy, and catalytic tests, the researchers at Jilin University show convincingly that the materials they prepare contain both mesoporosity and microporosity reminiscent of the properties of ZSM-5 crystals. Unfortunately, at this time there is no report describing the catalytic chemistry of the JLU-20 samples, and these cannot be compared to the two previous examples.

These three examples of mesoporous zeolites show the diversity of approaches that can be used to reach a common goal. Successful preparation of these complex materials depends on the self-assembly capabilities of the chosen nanomaterials as well as keen understanding by the researchers of both the aqueous chemistry of inorganic oxides and the colloidal chemistry of charged particles in an aqueous environment.

2.5 Synthesis of Extra-Large-Pore Zeolite ITQ-33

The group of Avelino Corma at the Institute de Tecnología Química (ITQ) de Valencia (see Site Reports-Europe on the International Assessment of Research in Catalysis by Nanostructured Materials website, www.wtec.org/private/catalysis; password required, contact WTEC) recently reported a new zeolite material with two important new structural characteristics (Corma et al., 2006). This material contains channels with 18-ring pores (Figure 2.8) ~12.5 Å in free diameter in one crystal direction, and in two perpendicular directions contains 10-ring (5.5 Å) channels. Prior to the

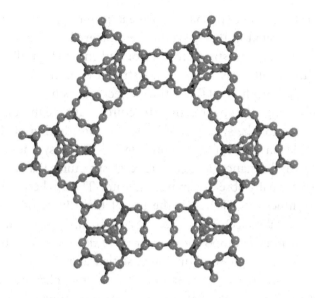

Fig. 2.8. Structure of the new 18-ring zeolite ITQ-33.

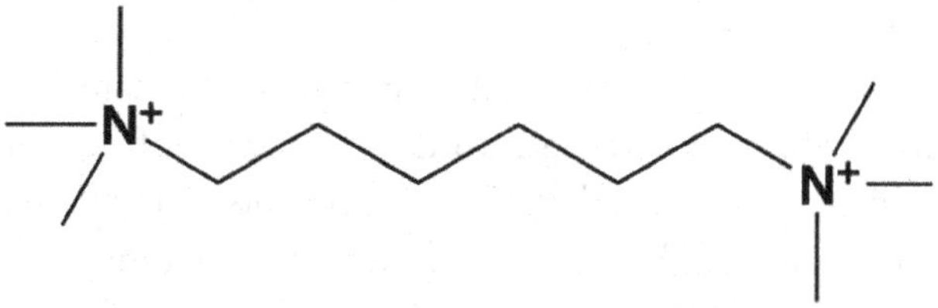

Fig. 2.9. Hexamethonium.

synthesis of ITQ-33, several extra-large-pore zeolites (containing more than 12 rings as the minimum pore dimension) had been synthesized. In particular, ECR-34, a gallium silicate, contains one-dimensional pores of similar size to ITQ-33. However, ITQ-33 has pores in three dimensions, and it is prepared with a simple structure director, hexamethonium cations.

The successful synthesis of ITQ-33 requires four ingredients. The first ingredient is a structure director (hexamethonium) that fills in the pore space not occupied by the silica framework (Figure 2.9). The second ingredient is the presence of some fluorine ions in the synthesis gel. These ions are incorporated in the as-synthesized material to balance (in part) the charge of the organic structure director and end up typically occluded in some of the small cages of the zeolite structure. In this form, the fluoride anion has both a stabilization effect for small cages and a structure-directing effect by selecting structures that have these cages. The third ingredient for the synthesis of ITQ-33 is the addition of some germanium oxide (in addition to silicon dioxide) to the synthesis gel. The larger Ge-O bond distances also stabilize cage structures that are different from the ones usually observed in pure silicates. The final crucial ingredient in the synthesis of ITQ-33 is the use of high-throughput synthesis methods. Originally, ITQ-33 was found as a small impurity in a set of exploratory synthesis compositions (Corma *et al.*, 2006) investigated by the Spanish group. By the combination of statistical design of experiments and high-throughput experimentation, Corma and collaborators found narrow synthesis conditions that allow for the preparation of this material in pure form. It is especially interesting that a small and flexible organic molecule such as hexamethonium is capable of stabilizing such a large and open structure.

The catalytic tests conducted with this zeolite indicate that the acidity of ITQ-33 is of similar strength to that of acidity of other high-silica zeolites. For instance, in the alkylation of benzene with propene to make cumene, ITQ-33 gives high selectivity at high conversions (with less than 0.01% by-products at 99% conversion), and the rate of deactivation is slower than the rate of deactivation of comparable commercial catalysts such as zeolite beta. ITQ-33 has also been tested for the cracking of vacuum oil, where it has shown several important qualities. First, it gives a conversion higher than zeolite beta and similar to zeolite USY. Second, it gives higher diesel selectivity without loss of the yield of butanes and propene. Because ITQ-33 can be prepared using a simple and inexpensive organic structure director, and because it has unique catalytic properties while maintaining good stability, it is quite possible that this rather exotic material will find industrial applications. This material is very new, and much can be expected from further study of its catalytic properties.

2.6 Heteropolyanions as Precursors for Desulfurization Catalysts

Hydrodesulfurization is a very important catalytic process used to remove sulfur from oil feedstocks by the selective hydrogenation of carbon-sulfur bonds. The catalysts are usually prepared by impregnation of γ-alumina with ammonium heptamolybdate and cobalt nitrate. This precursor material is sulfided to yield MoS_2 crystallites (a layered material) with cobalt atoms decorating the edges of the layers. It has been found empirically that a Co/Mo ratio of 0.5 gives the most active catalysts.

In order to improve over the existing catalysts, IFP and the Catalysis Laboratory at Lille in France (Martin *et al.*, 2005; Mazurelle *et al.*, 2008) have developed a new route to the preparation of hydrodesulfurization catalysts based on heteropolyanions (HPA) of molybdenum and cobalt. The basic idea is to use HPA with atomically mixed Co and Mo as a way to enhance the formation of small crystallites upon sulfidation. The starting point was the Anderson HPA of composition $CoMo_6O_{24}H_6(NH_4)_3$. Since the ratio of Co/Mo is below the optimal value, IFP and the Lamonier laboratory developed a synthesis of the dimer of this HPA $(Co_2Mo_{10}O_{38}H_4)^{6-}$ and then exchanged the ammonium form with cobalt to form a compound with the desired Co/Mo ratio.

Generally, the researchers found that the use of simple HPA (with a Co/Mo ratio different from the optimal) already showed improved activity over the materials prepared with heptamolybdate and cobalt nitrate. The new Co-exchanged Anderson HPA dimer resulted in even better activity than all previous samples. The origin of this enhanced activity can be explained based on TEM images of the standard catalysts and of the HPA-catalyst. The images show that the use of the HPA reduces the effective size (diameter) of the MoS_2 layers, generating more active sites per unit mass of catalyst. In the figure, disordered individual layers of the MoS_2 are the most commonly observed structure on the HPA-based catalysts. On the traditional catalysts, the MoS_2 layers are more organized in stacks of crystals, and in fact, the MoS_2 single layers can be observed to surround some of the alumina particles. The structure of the traditional catalysts leads to low levels of layer edges, the place where the active site of the catalysts is believed to be located.

2.7 Final Remarks

This chapter has described recent international advances in the synthesis of nanostructured catalysts that have shown promise to improve upon the activity or selectivity of existing catalysts. Multiple times in visits to labs in Europe and Asia, WTEC panelists observed the use of preformed nanostructures as starting blocks upon which the final structure is formed or developed. Panelists also observed that the characterization techniques developed for the nanotechnology revolution have been extremely helpful to the characterization of novel catalysts with nanostructures and have promoted an intensification of efforts across the world to improve important catalysts already in use or under development.

References

Bal, R., Tada, M., Sasaki, T., and Iwasawa, Y. (2006). Direct phenol synthesis by selective oxidation of benzene with molecular oxygen on an interstitial-N/Re cluster/zeolite catalyst, *Angew. Chem. Int. Ed.*, 45, pp. 448–452.

Choi, M., Cho, H. S., Srivastava, R., Venkatesan, C., Choi, D. H., and Ryoo, R. (2006). Amphiphilic organosilane-directed synthesis of crystalline zeolite with tunable mesoporosity, *Nat. Mater*, 5, pp. 718–723.

Christensen, C. H., Schmidt, I., Carlsson, A., Johannsen, K., and Herbst, K. (2005). Crystals in crystals: Nanocrystals within mesoporous zeolite single crystals, *J. Am. Chem. Soc.*, 127, pp. 8098–8102.

Christensen, C. H., Johannsen, K., Toernqvist, E., Schmidt, I., and Topsøe, H. (2007). Mesoporous zeolite single crystal catalysts: Diffusion and catalysis in hierarchical zeolites, *Catal. Today*, 128, pp. 117–122.

Corma, A., Díaz-Cabañas, M. J., Jordá, J. L., Martínez, C., and Moliner, M. (2006). High-throughput synthesis and catalytic properties of a molecular sieve with 18- and 10-member rings, *Nature*, 443, pp. 842–845.

DeSanto, P., Buttrey, D. J., Grasselli, R. K., Lugmair, C. G., Volpe, A. F., Toby, B. H., and Vogt, T. (2004). Structural aspects of the M1 and M2 phases in MoVNbTeO propane ammoxidation catalysts, *Zeitsch. Kristall.*, 219, pp. 152–165.

Grasselli, R. K., Burrington, J. D., Buttrey, D. J., DeSanto, P., Lugmair, C. G., Volpe, A. F., and Weingand, T. (2003). Multifunctionality of active centers in (amm)oxidation catalysts: From Bi–Mo–O_x to Mo–V–Nb–(Te, Sb)–O_x, *Topi. Catal.*, 23, pp. 5–22.

Janssen, A. H., Schmidt, I., Jacobsen, C. J. H., Koster, A. J., and De Jong, K. P. (2003). Exploratory study of mesopore templating with carbon during zeolite synthesis, *Micropor. Mesopor. Mater.*, 65, pp. 59–75.

Kusakari, T., Sasaki, T., and Iwasawa. Y. (2004). Selective oxidation of benzene to phenol with molecular oxygen on rhenium/zeolite catalysts, *Chem. Commun.*, pp. 992–993.

Kustova, M. Y., Hasselriis, P., and Christensen, C. H. (2004). Mesoporous MEL-type zeolite xingle crystal catalysts, *Catal. Lett.*, 96, pp. 205–211.

Lai, Z. P., Bonilla, G., Diaz, I., Nery, J. G., Sajauti, K., Amat, M. A., Kakkoli, E., Terasaki, O., Thomson, R. W., Tsapatsis, M., and Vlachos, D. G. (2003). Microstructural optimization of a zeolite membrane for organic vapor separations, *Science*, 300, pp. 456–460.

Lamonier, C., Martin, C., Mazurelle, J., Harlé, V., Guillaume, D., and Payen, E. (2007). Molybdocobaltate cobalt salts: New starting materials for hydrotreating catalysts, *Appl. Catal. B: Env.*, 70, pp. 548–556.

Li, D. F., Su, D. S., Song, J. W., Guan, X. Y., Hofmann, K., and Xiao, F. S. (2005). Highly steam-stable mesoporous silica assembled from preformed zeolite precursors at high temperatures, *J. Mater. Chem.*, 15, pp. 5063–5069.

Martin, C., Lamonier, C., Fournier, M., Mentré, O., Harlé, V., Guillaume, D., and Payen, E. (2005). Evidence and characterization of a new decamolybdocobaltate cobalt salt: An efficient precursor for hydrotreatment catalyst preparation, *Chem. Mater.*, 17, pp. 4438–4448.

Mazurelle, J., Lamonier, C., Lancelot, C., Payen, E., Pichon, C., and Guillaume, D. (2008). Use of the cobalt salt of the heteropolyanion $[Co_2Mo_{10}O_{38}H_4]^{6-}$ for the preparation of CoMo HDS catalysts supported on Al_2O_3, TiO_2 and ZrO_2, *Catal. Today*, 130, pp. 41–49.

McVicker, G. B., Kao, J. L., Ziemiak, J. J., Gates, W. E., Robbins, J. L., Treacy, M. M. J., Rice, S. B., Vanderspurt, T. H., Cross, V. R., and Ghosh, A. K. (1993). Effect of sulfur on the performance and on the particle size and location of platinum in Pt/KL hexane aromatization catalysts, *J. Catal.*, 139, pp. 48–61.

Sadakane, M., Watanabe, N., Katou, T., Nodasaka, Y., and Ueda, W. (2007). Crystalline Mo_3VO_x mixed-metal-oxide catalyst with trigonal symmetry, *Angew. Chem. Int. Ed.*, 46, pp. 1493–1496.

Srivastava, R., Choi, M., and Ryoo, R. (2006). Mesoporous materials with zeolite framework: Remarkable effect of the hierarchical structure for retardation of catalyst deactivation, *Chem. Commun.*, pp. 4489–4491.

Tada, M., Bal, R., Sasaki, T., Uemura, Y., Inada, Y., Tanaka, S., Nomura, M., and Iwasawa, Y. (2007). Novel re-cluster/HZSM-5 catalyst for highly selective phenol synthesis from benzene and O_2: Performance and reaction mechanism, *J. Phys. Chem. C*, 111, pp. 10095–10104.

Tang, T. D., Yin, C. Y., Wang, L. F., Ji, Y. Y., and Xiao, F. S. (2007). Superior performance in deep saturation of bulky aromatic pyrene over acidic mesoporous Beta zeolite-supported palladium catalyst, *J. Catal.*, 249, pp. 111–115.

Treacy, M. M. J. (1999). Pt agglomeration and entombment in single channel zeolites: Pt/LTL., *Micropor. Mesopor. Mater.*, 28, pp. 271–292.

Wang, L. F., Lin, K. F., Di, Y., Zhang, D. L., Li, C. J., Yang, Q., Yin, C. Y., Sun, Z. H., Jiang, D. Z., and Xiao, F. S. (2005). High-temperature synthesis of stable ordered mesoporous silica materials using mesoporous carbon as a hard template, *Micropor. Mesopor. Mater.*, 86, pp. 81–88.

3

Spectroscopic Characterization of Nanostructured Catalysts

Jeffrey T. Miller

3.1 Background

Much of the productivity improvements in the chemical industry within the last century could not have occurred without the help of catalysts. Catalytic materials provide improvements in reaction rate and product selectivity by lowering the activation energy by which chemical reactions occur. A wide variety of chemical substances show catalytic activity, however, most industrial processes employ heterogeneous catalysts that are comprised of metal, metal oxide, or metal sulfide nanosized particles on porous, high surface area supports. Additionally, some microporous and mesoporous catalysts have acidic properties. Three characteristics are common to every catalyst, e.g., activity, selectivity and stability. In order to optimize the catalyst efficiency, it is necessary to control the composition, physical properties and reactor environment. As a result, although the catalytic site may be of nanoscale size, characterization of the catalyst system requires methods that span the size range from atomic to nanometer to millimeter to meter length scales, Figure 3.1.

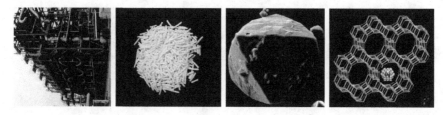

Fig. 3.1. Characterization of nanoscale catalysts at different length scales. From left to right: tens of meters (commercial reactor), millimeter scale (catalyst extrudates), micron scale (particle morphology), and atomic and nanometer scale (nanoparticles in mesoporous support) (courtesy of B. Weckhuysen, Utrecht University, Netherlands).

In addition, as the nature of the catalytic phase is often altered by the reaction environment, i.e., reacting to gases, temperature, time, etc., it is necessary to make measurements under realistic operating conditions. Many of the instrumental methods do not operate under extreme operating conditions; nevertheless, it remains a central goal in the catalysis field to make precise measurements of the catalyst under working conditions. This chapter reviews recent developments in the characterization of nanoparticle catalysts with particular emphasis on *in situ* or *operando* methods. This chapter only covers spectroscopic methods. Microscopy and related imaging techniques are covered separately in Chapter 4.

3.2 Laboratory Characterization Methods

3.2.1 Overview

Spectroscopic characterization methods are relied upon at all stages of the catalytic process, from preparation, pretreatment, and reaction to post-process analysis. Modern laboratories typically utilize a number of commercially available instruments to obtain this preliminary information. Typically available techniques include gas chromatography, physisorption and chemisorption, X-ray diffraction (XRD), scanning electron microscopy (SEM), transmission electron microscopy (TEM), electron spectroscopies (X-ray photo-electron [XPS] and Auger), and electron paramagnetic resonance (EPR), nuclear magnetic resonance (NMR) ultraviolet/visible (UV/Vis), Raman, and infrared (IR) spectroscopies. Across Asia (China,

Korea, and Japan) and Western Europe, most researchers that the WTEC panel visited had a significant number of new instruments with the latest improvements. This was especially true in China, where funding for new equipment was very high. In addition, many of the top universities in Asia and Europe had several faculty (4–10) involved with catalysis research, which allowed them to purchase larger, more expensive equipment, e.g., NMR, TEM, etc., for catalyst research.

With a few exceptions, most catalytic scientists are not instrument design engineers; thus, they rely on commercial instrument manufacturers for continued improvements. While instrument capabilities are improving at a rapid rate, standard commercial designs often do not allow for measurement under reaction conditions. As a result, catalytic scientists adapt existing equipment and develop controlled atmosphere reaction cells for *in situ* measurements. At many institutions in Europe, there are permanent staff scientists who maintain, operate, and modify equipment for the research scientists and graduate students. With years of experience, these highly skilled staff members are able to design special equipment and conduct difficult experiments necessary for the most in-depth and sophisticated analyses. In addition, these personnel provide the institution the continuity of skills necessary to more rapidly advance equipment development. In contrast, in the United States, once graduate students obtain their degrees and leave the labs, often critical skills are lost or must be relearned by the next generation of students.

3.2.2 *New spectroscopic capabilities and adaptations to standard laboratory instruments*

3.2.2.1 *In situ UV/Vis spectroscopy for zeolites*

Zeolites are crystalline, solid acid catalysts used for a variety of reactions. For large crystals, electron and optical microscopy show few internal details. UV/Vis spectroscopy can detect internal structure by detecting light-absorbing or -emitting hydrocarbons adsorbed in these zeolites (Karwacki *et al.*, 2007). Using an *in situ* cell to control the temperature and hydrocarbon partial pressure, the spacially resolved, time-dependent reactivity of different size reactants can be determined (Kox et al., 2007). Additionally, a laser source can reveal complementary fluorescence spectra (Stavitski et al., 2007). These methods have also been used to map the distribution of Brønsted acid sites in zeolites.

3.2.2.2 Catalyst preparation

Several new characterization techniques are being developed to follow the steps during catalyst preparation. In industry, catalyst preparation often involves the addition of a solution of metal salts to millimeter-sized pellets. Time-resolved Raman (Bergwerff et al., 2004; 2005), UV/Vis/NIR (Van de Water et al., 2005), and tomographic energy-dispersive (Beale et al., 2007) spectroscopies have been applied to follow the distribution and interaction of the metal salts on extrudates. Analysis of the Co and Mo addition to alumina, for example, shows that Co is uniformly distributed, Mo is distributed near the edge region, and there is a thicker edge region where a mixed Mo-Co complex deposits. In addition, there is a region near the center with low Mo concentration; thus, these analyses indicate an opportunity to better optimize the preparation for improved performance. Time-resolved Raman spectroscopy has also been used to observe the effect of metal ions in the synthesis of substituted aluminophosphate molecular sieves. Addition of Zn^{+2} ions, for example, alters the interaction between the gel and template, leading to a different template conformation and microporous structure (O'Brien et al., 2006).

In the absence of metal ions, hydroxyl groups on alumina, for example, can be detected by ^1H NMR, while the presence of metal ions quenches the signal. ^1H NMR imaging (MRI) has been developed to follow the distribution of metals on a single extrudate particle during impregnation (Bergwerff et al., 2007). Figure 3.2(a) shows the time-dependent impregnation of Co^{+2}-citrate complex on alumina at different times, and Figure 3.2(b) shows the different Co distributions that can be obtained by different preparation conditions. Depending on the catalyst application, one distribution will be more preferred than others.

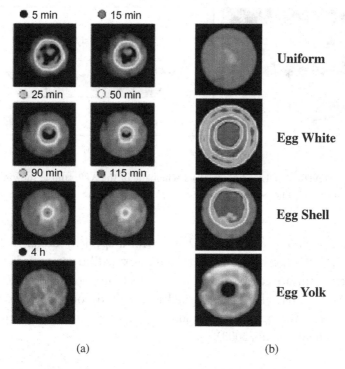

Fig. 3.2. (a) MRI of Co-citrate on alumina at different times; red indicates region of Co ions, (b) Distribution of Co on alumina with different preparation conditions, e.g., pH, Co:citrate ratio, etc. (courtesy of B. Weckhuysen, Utrecht University, Netherlands).

3.2.2.3 Catalyst surface-adsorbate spectroscopies

Critical to the understanding of any catalytic system are the structure, bonding, and electronic state of the active surface. Infrared and Raman spectroscopies have long been used to study the metal-adsorbate interaction at high temperatures in the gas phase.

3.2.2.4 Attenuated total reflectance infrared spectroscopy

Typically, IR spectroscopy is conducted in the gas phase, because high gas concentrations or liquids overwhelm the signal from the surface-adsorbed molecules. Using a transient Attenuated Total Reflectance (ATR) IR

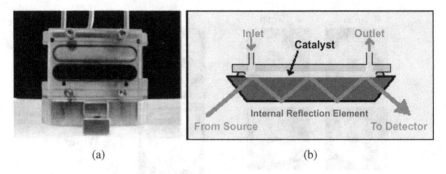

Fig. 3.3. (a) ATR IR cell for liquids, (b) a schematic of the ATR sample showing a thin oxide supported on a flat substrate (courtesy of A. Baiker, ETH, Switzerland).

spectroscopy (Bürgi and Baiker, 2006) with a specially designed IR cell (Figure 3.3), the species adsorbed on a thin oxide film supported on a flat IR reflecting surface can be determined in the liquid phase (Urakawa et al., 2003). ATR IR has also been applied to the study of supercritical CO_2 reactions, spectroscopy in flames, and spectroscopy in aqueous environments (Schneider et al., 2003).

3.2.2.5 Ultraviolet Raman spectroscopy

Raman spectroscopy, with its high resolution and wide spectral range, has been applied with success on heterogeneous catalysts for determination of adsorbed species and structure of crystalline and non-crystalline oxide phases. Raman is an especially attractive technique, due to its ability to characterize catalysts under reacting conditions. For many reactions, however, the catalyst is quickly covered with coke, which has a large fluorescence signal that interferes with the Raman bands. Utilization of an ultraviolet (UV) laser can minimize the fluorescence interference (Stair, 2002). The Dalian Institute of Chemical Physics has extensively used UV Raman spectroscopy to characterize the charge transfer of $p\pi$-$d\pi$ transitions from the metal ion to the oxide lattice in metal-containing microporous solids. This technique also can be used to identify and quantify the metal atoms in a zeolite structure (Li et al., 1999; Yu et al., 2000; Xiong et al., 2000). During the synthesis of Fe-ZSM-5, incorporation of Fe into the molecular sieve structure is followed by UV Raman.

3.2.2.6 *Spectroscopies for catalytic reactors*

In situ UV/Vis and Raman spectroscopies have been used to quantify the amount of coke formed during reaction (Nijhuis *et al.*, 2003; Tinnemans *et al.*, 2005). Using multiple detection points along the bed length of a catalytic reactor (Figure 3.4), it is possible to determine the amount of coke at different locations (Bennici *et al.*, 2007). During the reaction, the detectors monitor the buildup of coke, and at a predetermined level, the reaction is terminated and the catalyst is regenerated. Such detection and control systems may be applicable to larger industrial equipment.

3.2.2.7 *Magnetic resonance imaging of catalyst beds*

The advancement in magnetic resonance imaging (MRI) of catalyst systems is due in part to instrument advances, e.g., larger sample bores and more powerful magnets. The Magnetic Resonance Research Center

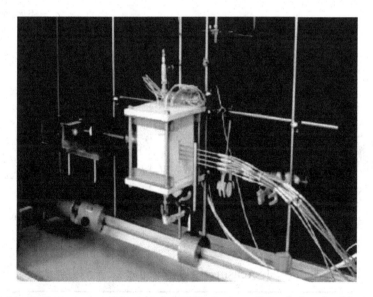

Fig. 3.4. Catalytic reactor with multiple UV/Vis detectors along the length of the catalyst bed (inside the furnace); at the top of the furnace is a Raman detector (courtesy of B. Weckhuysen, Utrecht University, Netherlands).

at Cambridge University has the latest instruments and is developing exciting and novel applications for catalyst applications (and other fields). For example, using ^{13}C MRI, the concentration of reactants and products as a position in the reactor can be quantitatively determined during reaction. In addition, the deactivation of the catalyst bed and formation of coke can also be followed. Other applications include determination of heterogeneity in catalyst particles and determination of diffusion coefficients in catalyst pellets during hydrogenation reactions.

MRI has also been applied to other systems, including the flow patterns in ceramic honeycomb catalytic converters (Gladden, 2003), the distribution of particles and voids in fluid beds as a function of gas rates (Figure 3.5), and liquid hold-up and wetting properties of liquid in trickle phase, fixed-bed reactors (Sederman and Gladden, 2001). The latter has been applied in commercial reactors.

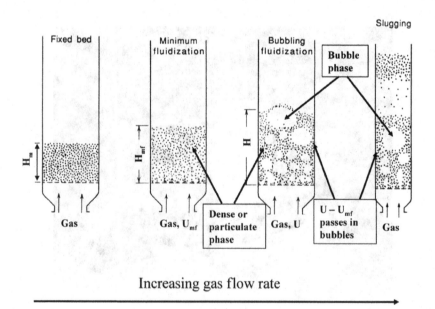

Fig. 3.5. MRI of fluid bed nanoparticle catalyst and gas voids with increasing gas velocity (courtesy of L. Gladden, Cambridge University, UK).

3.2.2.8 New instrument development

While most institutions and scientists rely on instrument manufacturers for advances in instrumentation capabilities, there is a significant effort in the Department of Chemical Physics at the Fritz Haber Institute in Germany to continually push the limits of experimental techniques by developing new instrumentation. For example, the department is currently developing an ultrahigh-resolution spectromicroscope (known as SMART), low-temperature scanning tunneling microscopy (STM), and a high-field W-band electron paramagnetic resonance spectrometer, as shown in Figure 3.6. SMART utilizes both chromatic and spherical aberration corrections to overcome the resolution limitations of currently available photoelectron emission microscopes (PEEM) to give lateral resolution of 2 nm and an energy resolution of 100 meV, which is claimed

Fig. 3.6. 95 GHz, ultrahigh-vacuum EPR spectrometer (courtesy of H. Freund, Fritz Haber Institute, Berlin, Germany).

to be the most ambitious project in the field of spectroscopic microscopy in the world. In addition to performing photoemission spectroscopy, X-ray photoemission electron microscopy (XPEEM), and low-energy electron microscopy (LEEM), the tool can also be used for NEXAFS (near edge X-ray absorption fine structure), XPS (X-ray photoelectron), and Auger spectroscopy, as well as for diffraction methods such as micro-spot low energy electron diffraction (LEED) and photoelectron diffraction (PED). The ultrahigh-vacuum (UHV), 95 GHz, high-field EPR will improve the spectral resolution compared to state-of-the-art instruments and has been used to study bimetallic nanoparticle catalysts, paramagnetic centers in Ziegler-Natta catalysts, and paramagnetic reaction intermediates. By combining low-temperature STM with photon STM, the group hopes to record Auger spectra with nearly atomic resolution.

In addition to the impressive new capabilities under development at the Fritz Haber Institute Department of Chemical Physics, its Department of Inorganic Chemistry also has an impressive capability of more conventional analytical tools:

- Spectroscopic methods: Raman (5 wavelengths available), *in situ* UV/Vis/NIR (2 units), *in situ* IR (4 spectrometers with DRIFTS and transmission capability)
- *In situ* photoelectron and synchrotron radiation methods
- Chemical and physical tools: elemental analysis, *in situ* XRD, variable atmosphere differential scanning calorimetry (DSC), calorimetry (4 units), adsorption, temperature programmed desorption (TPD), temperature programmed reduction (TPR), TEM (3 microscopes), and SEM (2 microscopes).

3.3 Synchrotron Methods

3.3.1 *Overview*

The increase in the use of synchrotron techniques for characterization of catalysis is, in part, due to the increased availability of synchrotrons, as well as to improved data quality and improvements software for data analysis. In addition, because of the penetrating depth of X-rays, synchrotron techniques can be applied to catalysts under reaction conditions.

The proliferation of synchrotron X-ray sources means that it is easier than ever to find a facility to do high-quality research.[1] In the last 15 years, three large new sources have been built for generating high-energy X-rays: the Super Photon Ring (SPring-8; Nishi Harima, Japan), the European Synchrotron Radiation Facility (ESRF; Grenoble, France), and the Advanced Photon Source (APS; Chicago, Illinois, USA), pictured in Figure 3.7.

Existence of these newer, so-called third-generation sources with higher-power X-ray sources means that experiments can be performed in much less time, at much lower concentrations, and with increased signal-to-noise. Several new facilities are under construction or in the planning stage in the UK (Diamond), France (Soleil), and the People's Republic of China (Shanghai Synchrotron Radiation Facility, SSRF).

3.3.2 Scattering techniques

X-ray diffraction has long been used for phase identification in heterogeneous catalysts, and with the high-intensity monochromatic light at synchrotron sources, diffraction patterns can be obtained in significantly shorter times. *In situ* X-ray diffraction patterns were obtained on a Pt on CeO_2-ZrO_2 auto exhaust catalyst in approximately 1 minute at the ESRF. By using a low dead-volume, *in situ*, capillary, plug-flow reactor with mass

Fig. 3.7. (*Left*) Third-generation synchrotrons Super Photon Ring-8 (Japan), (*right*) Advanced Photon Source (USA).

---

[1] Current synchrotron facilities throughout the world are listed at http://www-als.lbl.gov/als/synchrotron_sources.html.

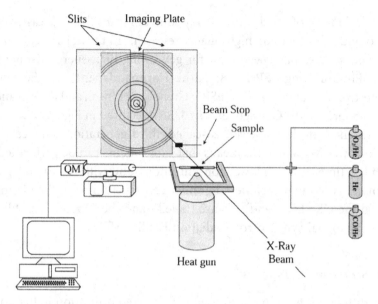

Fig. 3.8. *In situ*, capillary XRD reactor at the ESRF (courtesy of R. Psaro, Institute for Nanostructured Materials, Palermo, Italy).

spectrometer to detect the reaction products, the catalyst performance was monitored as XRD results were obtained and the reaction temperature increased (Figure 3.8). CO was found to reduce the CeO_2, and the phase transition from tetragonal to monoclinic was determined with different gas compositions and reaction temperatures (Martorana, 2003).

At the ESRF, time-resolved, *in situ* characterization of the synthesis of zinc-substituted microporous aluminophosphate was followed using multiple techniques, including small-angle X-ray scattering (SAXS), WAXS wide-angle X-ray scattering (WAXS), and XAFS. By combining these methods, the crystallization process could be simultaneously characterized at the molecular, nanoscopic and crystalline level (Fiddy *et al.*, 1999). Figure 3.9 shows a schematic representation and a photo of the beamline.

3.3.2.1 *X-ray absorption spectroscopy*

Since many nanoscale catalytic materials lack long-range order or are sufficiently small to be undetectable by XRD, X-ray absorption spectroscopy

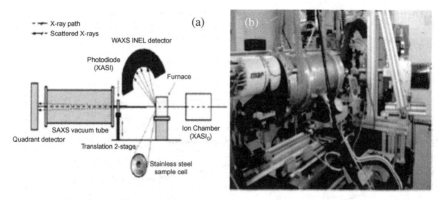

Fig. 3.9. ESRF's beamline for simultaneous measurement of SAXS, WAXS, and EXAFS (courtesy of B. Weckhuysen, Utrecht University, Netherlands).

(XAS) has become a widely used tool for structural (extended X-ray absorption fine structure, or EXAFS) and electronic (X-ray absorption near-edge structure, or XANES) characterizations. With the wider availability of higher-flux beamlines, EXAFS and XANES can be determined under reaction conditions to provide information on the active state of the catalyst. XAS characterizations also often provide complementary information to that determined by other methods.

With the high flux of the new, undulator beamlines at third-generation synchrotrons, data acquisition is significantly faster. Traditionally, spectra were taken by moving the monochrometer and collecting data for a few seconds, then moving to the next position. Such step-scan spectra typically took about one hour for a full spectrum. Additionally, at older facilities it was often necessary to average several spectra to have sufficient signal to noise to resolve subtle structural features. At new facilities with the higher flux, it is no longer necessary to stop the monochrometer to acquire data. In the continuous-scan or quick-EXAFS mode, data are obtained in a few minutes with excellent signal to noise. Step-scan spectra are now used for very dilute samples, although these may also be obtained in fluorescence. Very fast spectra can also be obtained by a dispersive EXAFS beamline design. With the dispersive EXAFS beamline, a bent crystal gives all energies at one time, allowing for very fast spectra

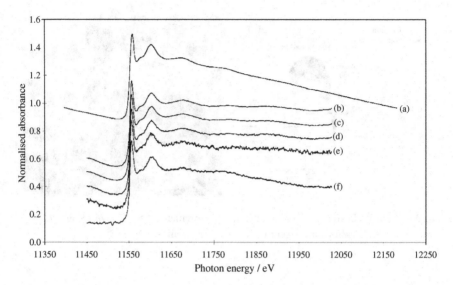

Fig. 3.10. EXAFS spectra of 5% Pt on silica catalyst: (a) step-scan spectrum taken in 45 min.; (b–f) dispersive EXAFS, spectra taken at (b) 50 sec, (c) 5 sec, (d) 0.5 sec, (e) 0.05 sec, and (f) dispersive EXAFS spectrum of 1% Pt on silica taken in 21 sec (Fiddy et al., 1999).

acquisition. The ESRF (France) and Photon Factory (Japan) have dispersive EXAFS beamlines where spectra can be obtained in less than 0.1 sec (Figure 3.10).

3.3.2.2 High-energy resolution XANES

XANES spectroscopy probes the local geometry and the oxidation state of the absorbing atom. XANES is also sensitive to adsorbates at the nanoparticle surface (Oudenhuijzen et al., 2005; Ramaker et al., 2003; Ramaker and Koningsberger, 2002; Ankudinov et al., 2002). The intensity of the first feature in the L_3 edge spectrum reveals the number of holes in the d-band and, therefore, reflects charge transfer after adsorption of molecules. Fluorescence XANES, with an instrumental broadening below the core hole lifetime, greatly enhances the spectral resolution (Glatzel and Bergmann, 2005; Hämäläinen et al., 1991; de Groot et al., 2002; de Groot, 2005). Figure 3.11(a) illustrates the enhancement in XANES resolution of

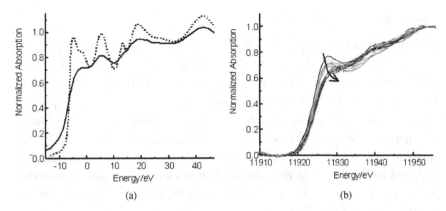

Fig. 3.11. (a) XANES of Au foil: solid line = transmission; dotted line = high-energy fluorescence, (b) Time-resolved (2 sec) XANES spectra of oxidized Au/alumina catalyst in 1% CO (van Bokhoven et al., 2006).

Au foil measured in two modes: normal transmission (solid line) and high energy-resolution fluorescence (dotted line). A small white line is visible in the transmission spectrum while the XANES is enhanced in the high-energy resolution fluorescence spectrum. In addition, in the high-energy resolution spectrum, all other features are much more pronounced (van Bokhoven et al., 2006).

The XANES of an oxidized Au/Al_2O_3 catalyst shows increased white-line intensity consistent with partial oxidation of the Au surface. Figure 3.11(b) shows the evolution of the XANES spectra after switching to 1% CO. The spectra were recorded every 2 seconds and indicate that reduction is very fast, much faster than Au oxidation (van Bokhoven et al., 2006). In addition, upon reduction, small amounts of CO_2 are observed in the mass spectrometer, suggesting that catalytic site is an oxidized Au surface on the metallic particle.

The increased sensitivity of the high-energy resolution fluorescence XANES combined with the *in situ*, time-resolved fast data acquisition provides a powerful new tool for quantitatively determining the kinetics of the elementary steps in the catalytic cycle. Coupled with the ability to simultaneously measure the reaction rate, it should be possible to determine how the catalyst composition (nanoparticle supports and modifiers) alter the kinetics of these elementary steps in the catalytic cycle.

3.3.2.3 Spacially resolved XAS

XAS spectroscopy is typically a bulk technique, i.e., the spectrum is an average of all species that are present. However, during preparation and pretreatment, several phases may be produced. Thus, the EXAFS and XANES spectra may represent both the catalytic and non-catalytic species. XANES spectra of oxidized and reduced metals, however, occur at different energies and have different shapes. Therefore, the fraction of each in a partially reduced catalyst can be determined. With an *in situ* reaction cell designed to hold a single 600 nm Fe_2O_3 particle, the fraction of oxidized and reduced Fe during reduction and during the Fischer–Tropsch reaction was determined by Fe XANES.

3.3.3 Millibar XPS at BESSY synchrotron (Germany)

In a 10-year collaboration between M. Salmeron of the Lawrence Berkeley Laboratory, R. Schlögl of the Fritz Haber Institute, and personnel at the BESSY (Berliner Elektronenspeicherring-Gesellschaft für Synchrotronstrahlung m.b.H., Germany) synchrotron, an *in situ* XPS beamline ("Innovative Station for *In Situ* Spectroscopy," or ISISS) was developed that could operate with pressures up to 10 mbar. The beamline exploits the synchrotron radiation of a bending magnet and is optimized to deliver photons in the energy range between 80 eV and 2000 eV with a high photon flux. It can be operated in different modes, that is, high flux, high spectral resolution, or higher order suppression (thereby increasing the spectral purity of the X-ray beam). The installation of a ventilation and safety system makes it feasible to use hazardous substances (e.g., flammable, explosive, or poisonous) as feed gas. In addition, ISISS is also equipped with a mass spectrometer to measure the reaction products while the spectrum is being obtained. A schematic of the high-pressure XPS and a photo of the experimental station are shown in Figure 3.12 (FHI n.d.).

The high-pressure XPS at BESSY has been used to study the oxygen species on Ru, Ag, and Cu. For example, subsurface oxygen was found in polycrystalline Cu foil during the partial oxidation of methanol to

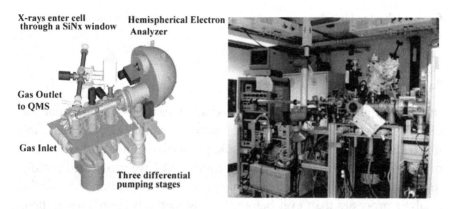

Fig. 3.12. A schematic (*left*) and photo (*right*) of the HP-XPS at ISISS beamline at BESSY (Germany) (courtesy of Fritz Haber Institute).

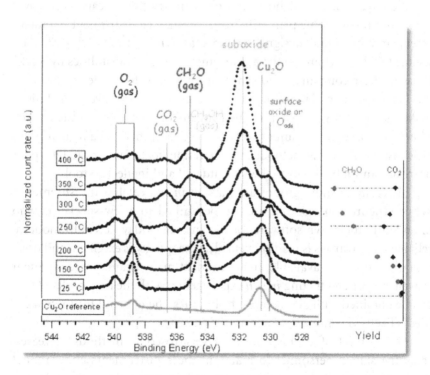

Fig. 3.13. Photoemission spectra of a polycrystalline Cu foil under reaction condition: $CH_3OH/O_2 = 3:1$ and 0.4 mbar (Knop-Gericke n.d.).

formaldehyde. The abundance of this species is correlated to the yield of formaldehyde, as Figure 3.13 shows.

3.4 Conclusions

In-depth, high-quality spectroscopic characterizations are essential for effective development of improved catalyst technology. New tools for understanding the structure, electronic properties, and adsorbed species of nanoscale catalysts under reaction conditions have progressed rapidly in the last decade. Today, there are many more methods to understand the catalytic processes than ever before. As new methods are reported, these are quickly adapted by other groups and applied to other materials. Often new approaches are also being adapted from other fields such as biology and physics. While there remains the need for higher resolution and quantification with many methods, improved instrumental capabilities will likely continue to come primarily from commercial instrument manufacturers and beamline design engineers at synchrotrons. The catalytic scientist will contribute to the advancement of these capabilities by working with these companies and synchrotron groups to define the needs of the field. For many laboratory instruments, these improved capabilities make the current generation of instruments obsolete in about ten years. Thus, it becomes important to replace aging equipment on a regular basis. New insights into new methods of preparation, materials, or catalytic phenomena cannot be discovered with outdated and ineffective tools.

Increasingly, more groups are making *in situ* catalyst measurements. While synchrotron research used to be limited to a few specialists who often wrote their own software for the analysis, today, access to modern, reliable synchrotrons is readily available, data quality is excellent, and user-friendly software is available. In addition, at many beamlines, equipment is available for conducting *in situ* experiments. With the level of detail that can be obtained with synchrotron techniques, the demand for these facilities is likely to continue to increase.

As the level of sophistication of spectroscopic methods increases, retaining skilled personnel to maintain these instruments and conduct difficult experiments becomes important. In organizations where there is no permanent staff support, once students graduate, expertise is lost and must be relearned; thus, the amount of time available for advancing the methodologies is reduced. At synchrotrons, most beamlines have staff scientists who assist with the experimental setup and execution of the experimental plan. They also maintain the beamline calibrating instruments, performing routine maintenance, and upgrading the equipment. Many universities in Europe have also recognized the value of permanent staff for operating multiuser instruments. They conduct instrument training, perform maintenance and upgrades, build sample cells, and assist new users with experiments. In addition to improving the quality of data and experiments, these staff scientists provide the continuity in expertise upon which further advancements can be built. This is a valuable practice.

References

Ankudinov, A. L., Rehr, J. J., Low, J. J., and Bare, S. R. (2002). Reply to Ramaker and Koningsberger, *Phys. Rev. Lett.*, 89, p. 139702.

Beale, A. M., Jacques, S. D. M., Bergwerff, J. A., Barnes, P., and Weckhuysen, B. M. (2007). Tomographic energy dispersive diffraction imaging as a tool to profile in three dimensions the distribution and composition of metal oxide species in catalyst bodies, *Angew. Chem. Int. Ed.*, 46, pp. 8832–8835.

Bennici, S. M., Vogelaar, B. M., Alexander, T., Nijhuis, and Weckhuysen, B. M. (2007). Real-time control of a catalytic solid in a fixed-bed reactor based on in situ spectroscopy, *Angew. Chem. Int. Ed.*, 46, pp. 5412–5416.

Bergwerff, J. A., Lysova, A. A., Espinosa Alonso, L., Koptyug, I. V., and Weckhuysen, B. M. (2007). Probing the transport of paramagnetic complexes inside catalyst bodies in a quantitative manner by magnetic resonance imaging, *Angew. Chem. Int. Ed.*, 46, pp. 7224–7227.

Bergwerff, J. A., van de Water, L. G. A., Visser, T., de Peinder, P., Leliveld, B. R. G., de Jong, K. P., and Weckhuysen, B. M. (2005). Spatially resolved Raman and UV/Visible-NIR spectroscopy on the preparation of supported catalyst bodies: Controlling the formation of H2PMo11CoO405- inside Al_2O_3 pellets during impregnation, *Chem. Eur. J.*, 11, pp. 4591–4601.

Bergwerff, J. A., Visser, T., Leliveld, B. R. G., Rossenaar, B. D., de Jong, K. P., and Weckhuysen, B. M. (2004). Envisaging the physicochemical processes during the preparation of supported catalysts: Raman microscopy on the impregnation of Mo onto Al_2O_3 extrudates, *J. Am. Chem. Soc.*, 126, pp. 14548–14556.

Bürgi, T., and Baiker, A. (2006). Attenuated total reflection infrared spectroscopy of solid catalysts functioning in the presence of liquid-phase reactants, *Adv. Catal.*, 50, pp. 227–283.

de Groot, F. M. F. (2005). Multiplet effects in X-ray spectroscopy, *Coord. Chem. Rev.*, 249, pp. 31–63.

de Groot, F. M. F., Krish, K. H., and Vogel, J. (2002). Spectral sharpening of the Pt L edges by high-resolution X-ray emission, *Phys. Rev. B.*, 66, pp. 195112–19117.

de Groot, F. M. F., and Weckhuysen, B. (n.d.) Utrecht University, Netherlands, unpublished results.

Fiddy, S. G., Newton, M. A., Dent, A. J., Salvini, G., Corker, J. M., Turin, S., Campbell, T., and Evans, J. (1999). In situ energy dispersive EXAFS (EDE) of low loaded $Pt(AcAc)_2/H_I\ SiO_2$ catalyst precursors on a timescale of seconds and below, *Chem. Commun.*, pp. 851–852.

Fritz Haber Institute of the Max Planck Institute (n.d.) "High pressure XPS / XAS Spectrometer at BESSY". http://w3.rz-berlin.mpg.de/ac/surfana/xps.pdf.

Gladden, L. F. (2003). Recent advances in MRI studies of chemical reactors: Ultrafast imaging of multiphase flows, *Top. Catal.*, 24, pp. 19–28.

Glatzel, P., and Bergmann, U. (2005). High resolution 1s core hole X-ray spectroscopy in 3d transition metal complexes: Electronic and structural information, *Coord. Chem. Rev.*, 249, pp. 65–95.

Hämäläinen, K., Siddons, D. P., Hastings, J. B., and Berman, L. E. (1991). Elimination of the inner-shell lifetime broadening in X-ray absorption spectroscopy, *Phys. Rev. Lett.*, 67, pp. 2850–2853.

Karwacki, L., Stavitski, E., Kox, M. H. F., Kornatowski, J., and Weckhuysen, B. M. (2007). Intergrowth structure of zeolite crystals as determined by optical and fluorescence microscopy of the template-removal process, *Angew. Chem. Int. Ed.*, 46, pp. 7228–7231.

Knop-Gericke, A. (n.d.) Department of Inorganic Chemistry, Fritz Haber Institute of the Max Plank Society, Berlin, Germany. http://w3.rz-berlin.mpg.de/ac/surfana/highlights.html.

Kox, H. H. F., Stavitski, E., and Weckhuysen, B. M. (2007). Non-uniform catalytic behavior of zeolite crystals as revealed by in situ optical microscopy, *Angew. Chem. Int. Ed.*, 46, pp. 3652–3655.

Li, C., Xiong, G., Xin, Q., Liu, J.-K., Ying, P.-L., Feng, Z.-C., Li, J., Yang, W.-B., Wang, Y.-Z., Wang, G.-R., Liu, X.-Y., Lin, M., Wang, X.-Q., and Min, E.-Z. (1999). UV resonance Raman spectroscopic identification of titanium atoms in the framework of TS-1 zeolite, *Angew. Chem. Int. Engl. Ed.*, 38, pp. 2220–2222.

Martorana, A., Deganello, G., Longo, A., Deganello, F., Liotta, L., Macaluso, A., Pantaleo, G., Balerna, A., Meneghini, C., and Mobillio, S. (2003). Time-resolved

X-ray powder diffraction on a three-way catalyst at the GILDA beamline, *J. Synchrotron Rad.*, 10, pp. 177–182.

Nijhuis, T. A., Tinnemans, S. J., Visser, T., and Weckhuysen, B. M. (2003). Operando spectroscopic investigation of supported metal oxide catalysts by combined time-resolved UV-Vis/Raman on-line mass spectrometry, *Phys. Chem. Chem. Phys.*, 5, pp. 4361–4365.

O'Brien, M. G., Beale, A. M., Catlow, C. R. A., and Weckhuysen, B. M. (2006). Unique organic-inorganic interactions leading to a structure-directed microporous aluminophosphate crystallization as observed with *in situ* Raman spectroscopy, *J. Am. Chem. Soc.*, 128, pp. 11744–11745.

Oudenhuijzen, M. K., van Bokhoven, J. A., Miller, J. T., Ramaker, D. E., and Koningsberger, D. C. (2005). Three-site model for hydrogen adsorption on supported platinum particles: Influence of support ionicity and particle size on the hydrogen coverage, *J. Am. Chem. Soc.*, 127, pp. 1530–1540.

Ramaker, D. E., and Koningsberger, D. C. (2002). Comment on effect of hydrogen adsorption on the X-ray absorption spectra of small Pt cluster, *Phys. Rev. Lett.*, 89, p. 139701.

Ramaker, D. E., Teliska, M., Zhang, Y., Stakheev, A. Y., and Koningsberger, D. C. (2003). Understanding the influence of support alkalinity on the hydrogen and oxygen chemisorption properties of Pt particles, *Phys. Chem. Chem. Phys.*, 5, pp. 4492–4501.

Schneider, M. S., Grunwaldt, J.-D., Bürgi, T., and Baiker, A. (2003). High pressure view-cell for simultaneous in situ infrared spectroscopy and phase behavior monitoring of multiphase chemical reactions, *Rev. Sci. Instrum.*, 74, pp. 4121–4128.

Sederman, A. J., and Gladden, L. F. (2001). Magnetic resonance visualization of single- and two-phase flow in porous media, *Magn. Reson. Imag.*, 19, pp. 339–343.

Sederman, A. J., and Gladden, L. F. (2002). Magnetic resonance imaging as a quantitative probe of gas-liquid distribution and wetting efficiency in trickle-bed reactors, *Chem. Eng. Sci.*, 56, pp. 2615–2628.

Stair, P. C. (2002). *In-situ Spectroscopy in Heterogeneous Catalysis*, ed. Haw, J. F., "In situ ultraviolet Raman spectroscopy", (Wiley-VCH), pp. 121–138.

Stavitski, E., Kox, H. H. F., and Weckhuysen, B. M. (2007). Revealing shape selectivity and catalytic activity trends within the pores of H-ZSM-5 crystals by time- and space-resolved optical and fluorescence micro-spectroscopy, *Chem. Eur. J.*, 13, pp. 7057–7065.

Tinnemans, S. J., Kox, M. H. F., Nijhuis, T. A., Visser, T., and Weckhuysen, B. M. (2005). Real time quantitative Raman spectroscopy of supported metal oxide catalysts without the need of an internal standard, *Phys. Chem. Chem. Phys.*, 7, pp. 211–216.

Urakawa, A., Wirz, R., Bürgi, T., and Baiker, A. (2003). ATR-IR flow-through cell for concentration modulation excitation spectroscopy: Diffusion experiments and simulations, *J. Phys. Chem. B*, 107, pp. 13061–13068.

van Bokhoven, J. A., Louis, C., Miller, J. T., Tromp, M., Safonova, O. V., and Glatzel, P. (2006). Activation of oxygen on gold-alumina catalysts: In situ high energy-resolution and time-resolved X-ray spectroscopy, *Angew. Chem. Int. Ed.*, 45, pp. 4651–4654.

van de Water, L. G. A., Bergwerff, J. A., Nijhuis, T. A., de Jong, K. P., and Weckhuysen, B. M. (2005). UV-Vis microspectroscopy: Probing the initial stages of supported metal oxide catalyst preparation, *J. Am. Chem. Soc.*, 127, pp. 5024–5025.

Xiong, G., Li, C., Li, H.-Y., Xin, Q., and Feng, Z.-C. (2000). Direct spectroscopic evidence for vanadium species in V-MCM-41 molecular sieve by UV resonance Raman spectroscopy, *Chem. Commun.*, pp. 677–678.

Yu, Y., Xiong, G., Li, C., and Xiao, F.-S. (2000). Characterization of iron in the framework of MFI-type zeolites by UV resonance Raman spectroscopy, *J. Catal.*, 194, pp. 487–490.

4

Electron and Tunneling Microscopy of Nanostructured Catalysts

Renu Sharma

4.1 Introduction

The overall performance (that is, activity and selectivity) of catalysts is controlled by their size, morphology, nanostructure, and interaction with the support. Whereas nanometer-sized particles generally have higher reactivity due to their relatively high surface area, their morphology determines if they are bound by the active surfaces. For bimetallic, binary, and ternary oxide catalysts, the inter-granular and/or intragranular heterogeneity in the sample may be a critical factor that determines their performance. Last but not least, since catalyst nanoparticles are generally supported on high-surface-area supports such as carbon, alumina, silica, or titania, we need to understand the interaction of the particle with the support, because the structure and morphology of the catalysts are strongly influenced by the nature of the support surface.

First and foremost the morphology, structure and chemistry of nanostructured catalysts are directly controlled by the synthesis. Therefore the ultimate goal of the characterization is to relate the synthesis methods to the structure and morphology and thereby to the activity and selectivity (performance) for heterogeneous catalysis at the nanoscale. Secondly, the temperature, pressure, and chemical environment often affect the morphology, structure and composition of the catalysts during operation, which may result in reduced activity, often known as catalyst poisoning. Therefore catalyst samples are characterized at various levels during synthesis and operation. Bulk characterization techniques, as described in Chapter 3, can often provide adequate information to achieve this goal. However, in order to understand atomic-level changes occurring during the synthesis and functioning of the catalyst, it is necessary to characterize them at high spatial (lateral) resolution. Therefore electron and tunneling microscopy techniques are often employed to determine the structural and chemical variations that exist in as-prepared or -used catalysts.

4.1.1 Overview of high resolution characterization techniques

Nanoscale characterization is employed for (a) general characterization of catalysts, and/or (b) to understand the changes occurring during the synthesis and/or functioning of catalysts. The latter is achieved either by characterization of materials before and after the reaction or by directly observing the dynamic process under reactive environments (*in situ*). There are several electron and tunneling microscopy techniques that can be successfully employed to obtain nanoscale information from individual catalyst and support particles. Following is a short description of some of these techniques.

4.1.1.1 Scanning probe microscopy (SPM)

Scanning probe microscopy (SPM) is a general term employed for microscopy techniques such as scanning tunneling microscopy (STM) and atomic force microscopy (AFM) that are used to form images of surfaces using a physical probe that scans the specimen. In simple terms STM is based on the theory that when a conducting tip is brought near a metallic or semiconducting surface (within 0.3 to 1 nm distance), a current flow that is dependent on the local density of states is observed. The magnitude of this current depends upon the gap and the local barrier height. This current, also called tunneling current, represents the surface structure. Therefore, STM is very suitable for obtaining atomic-level structural and topographic information about the surfaces. Recent advances in instrumentation allow researchers to obtain atomic-level changes occurring on the catalyst surfaces in liquid or gaseous environments at different temperatures. Time- and temperature-resolved images are used to observe and understand the surface reconstruction and to identify surface adsorbates and chemical processes occurring during catalysis.

4.1.1.2 *Scanning electron microscopy (SEM)*

Scanning electron microscopy (SEM) is widely employed to obtain topographic images and chemical information. In this case, as an electron beam (5–30 KeV) is used to scan the sample, a number of signals are generated, including secondary electrons, back-scattered electrons, and X-rays. Both secondary and back-scattered electrons are used to form images, and chemical information is obtained from the X-ray signals using energy-dispersive or wavelength-dispersive spectrometers (EDS or WDS, respectively). The spatial resolution for chemical analysis is dependent upon the probe size (electron gun and probe-forming optics) and can vary from a few nanometers to several hundred nanometers. Secondary electrons are emitted from the atoms occupying the sample surface and produce surface images, whereas backscattered electrons are produced by the reflection of the primary beam, and therefore their intensity depends upon the atomic number. The latter images are most suitable for locating heavy metal particles on light oxide surfaces. Environmental SEM (ESEM) with heating and cooling stages is often used to observe catalysts under working conditions. Recently, orientation image microscopy (OIM) has been incorporated with SEM and is used to obtain the preferred orientation of various particles and their relationship to the support.

4.1.1.3 *Scanning/transmission electron microscopy (STEM/TEM)*

Transmission electron microscopy (TEM), high-resolution TEM (HRTEM), and scanning-transmission electron microscopy (STEM) are

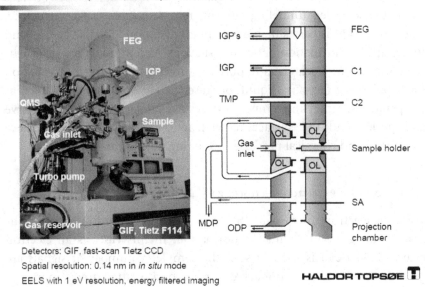

Fig. 4.1. HRTEM instrument (*left*) and cross-section of the column (*right*) showing the electron source (labeled FEG), condenser lenses (C1 and C2), objective lenses (OL), location of sample, and projection chamber where the image is formed. Components of the modification made to introduce gases in the sample chamber area (gas inlet), and the differential pumping systems (MDP, ODP, and IGP) are also shown (courtesy of Haldor Topsøe).

more complicated and expensive techniques compared to STM and SEM. Here, high-energy (typically 100–1000 KeV) electrons are used to obtain diffraction data and images from very thin (10–50 nm) samples. Figure 4.1 shows an HRTEM instrument (left) and a simple cross-section (right) of a TEM, including the modifications incorporated for observation of environmental effects, generally known as an environmental transmission electron microscope (ETEM). In the HRTEM, high-energy electrons (of short wavelength) generated by a field-emission gun (FEG) are focused on the sample by magnetic lenses (C1 and C2 in Figure 4.1). The electrons interact with the electron potential (density of states) of the atoms as they travel through the sample. An objective lens (OL), also known as an image-forming lens, is used to focus the transmitted and the diffracted beams to form diffraction-pattern and high-resolution images that are magnified and projected by another set of lenses. Some of the electrons lose energy as they interact with the sample and generate an X-ray signal (similar to SEM). Both the X-ray signal and the energy lost by electrons are characteristic of the electron potential in the sample and can be collected using suitable detectors, e.g., EDS and electron energy-loss spectra (EELS), to obtain local chemical information.

STEM images are formed by scanning the sample (similar to SEM) using a nanometer-size probe. By using suitable detectors, such as a high-angle annular dark-field (HAADF) detector, atomic-resolution images are collected, with intensity in the image directly proportional to the square of scattering power of the atoms (Z^2). HAADF images are most suitable for obtaining the size distribution of catalyst particles. Since most of catalyst particles are heavy metals (Au, Pt, Ni, Fe, Cu, MoS_2, etc.) supported on lightweight oxide supports such as silica and alumina, very-high-contrast images are obtained. Although HAADF images are not as easy to obtain as SEM or TEM images, particle size can be determined unambiguously, because the boundary of the particles is clearly marked due to the high contrast in the images. Moreover, because these images are usually obtained using very small probe size (0.2–0.5 nm), nanoscale chemical information, obtained using either EDS or EELS, can be combined with the images to obtain the chemical information from the regions around and within the catalyst particles. Therefore, STEM analysis is often employed to characterize nanoscale heterogeneity in the samples.

In order to observe atomic-level structural and chemical changes, the TEM column can be modified to incorporate a gas inlet system and differential pumping apertures to constrain reactive gases in the sample area (Figure 4.1, right). With careful design, the rest of the column can be kept at high vacuum while the sample area can be at up to 50 mbar gas pressure. Suitable sample heating holders are used to observe the dynamic evolution of nanostructures and chemical changes at the atomic scale in gaseous environments at elevated temperatures (900–1000°C). Both TEM and STEM images can be obtained using an ETEM (Gai, 1999; Sharma, 2005).

4.1.1.4 Low energy electron microscopy (LEEM)

As the name implies, low-energy electron microscopy (LEEM) uses very-low-energy electrons (0–100 eV) to form images based on the diffraction and phase of the sample. It is a surface characterization technique, because these electrons do not have enough energy to penetrate through the sample. The image formation does not require scanning; therefore, it is fast and often used to observe time-resolved changes.

4.1.1.5 Photo emission electron microscopy (PEEM)

In the original photo emission electron microscope, the photons from UV light incident on the sample were used to form images but now other light sources are also being used. The principle of image formation is that the photoelectron emission is produced if the energy of the incident photons on a sample is larger than its work function. An electron-optical imaging system is used to extract the photo-emitted electrons by a large electric field that is applied between the sample and the first electrode of the electron optical system. A combination of several such systems is used to obtain a full-field image of the emitted electrons onto a detector such as a phosphor that converts electrons into visible light. Several other photon sources, such as X-rays, are currently being used to form images that can also provide chemical information.

WTEC panelists found that SPM, SEM, and STEM/TEM are the techniques most commonly employed in the various laboratories we visited in Asia and Europe. Therefore, this chapter focuses on those techniques only.

4.2 General Characterization of Catalyst Particles

Particle size, shape, and composition (in the case of bimetallic or mixed oxide catalysts) can be determined using STM, SEM, TEM, and/or STEM. These techniques provide information at various levels of image, spatial, and temporal resolution. SEM is often employed to quickly scan the samples synthesized under different conditions or using different methods. The general information about the particle size distribution and chemical composition of individual particles thus obtained is then used to optimize the synthesis process for a specific application.

An example is the characterization of titania (TiO_2), which is an important photocatalyst with applications ranging from water purification to synthesis of hydrogen by water splitting as well as a catalyst support material. It has been shown that catalytic properties of TiO_2 can be improved by changing its morphology. Its photocatalytic activity is limited by the band gap (3.0–3.2 eV for anatase). Professor Centi (University of Messina, Italy, and member of the EU Integrated Design of Catalytic Nanomaterials consortium) has shown that ordered helical structures (Figure 4.2) produced by the anodization of titanium foils have improved properties in the UV-visible region. Titania nanostructured thin films (TNT) were observed to have a linear relationship between wall thickness and the band gap. SEM images (Figure 4.2) were used to optimize the

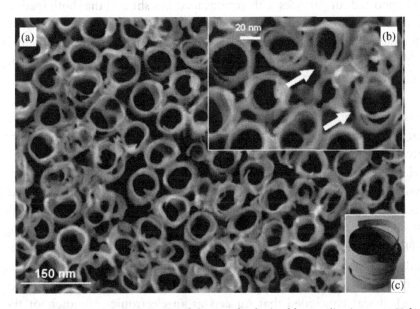

Fig. 4.2. Field-emission SEM image of the sample obtained by anodization at 15 V for 45 minutes. In the inset (a, b), an expansion of some TNT shows the nanostructure of a helical nanocoil. In the inset (c), a photo of a rolled paper streamer is reported as an example to better illustrate the nanostructure (Perathoner et al., 2007).

synthesis methods that will produce the TNTs with desired morphology and thereby improved properties (Perathoner et al., 2007).

Particle size is another crucial parameter for catalytic activity, especially for Au catalysts. Au is normally an inert material, but in 1995 Hutchings and colleagues at Cardiff University in the UK (see Site Reports-Europe on the International Assessment of Research in Catalysis by Nanostructured Materials website, www.wtec.org/private/catalysis; password required, contact WTEC) showed that nanostructured Au particles with less than 55 atoms are the most active catalyst for a higher alkene epoxidation reaction; however, due to the high mobility of Au atoms, it is difficult to synthesize and control their size on a support (Hughes et al., 2005). Dr. Lambert's group at Cambridge University (see Site Reports-Europe on the International Assessment of Research in Catalysis by Nanostructured Materials website, www.wtec.org/private/catalysis; password required, contact WTEC) has synthesized Au nanoparticles on various supports using cluster chemistry. TEM analysis of supported and unsupported Au particles with various loadings showed that both loading and support control the particle size (Figure 4.3). Higher selectivity of styrene oxidation was achieved using small particles (≈ 1.5 nm), obtained with 0.6% loading on boron nitride (BN) and silica (SiO_2) and conversion rates of 19.2% and 25.8%, respectively.

Hydrogen peroxide (H_2O_2) is a key chemical commodity with numerous applications; its direct synthesis is one of the most challenging targets for the control of reaction selectivity in heterogeneous catalysis. A major breakthrough for direct synthesis was obtained using an Au catalyst (Hughes et al., 2005; Enache et al., 2006). Later, a comparative study of the synthesis of H_2O_2 using 5% loading of Au, Au/Pd (4 + 1% and 2.5 + 2.5%), and Pd on TiO_2 showed that maximum productivity is obtained with Au–Pd catalyst (equal amounts Au and Pd) (Hutchings, 2007). Using STEM imaging combined with nanoscale EDS, detailed analysis of the individual particles in the most reactive samples (Figure 4.4) showed that particles consist of a Pd layer surrounding an Au core (core-shell structure). It was concluded that Au acts as an electronic promoter for the Pd-rich surface.

These examples illustrate the need for nanoscale characterization of both catalyst and support to understand and improve the activity and

Electron and Tunneling Microscopy of Nanostructured Catalysts 73

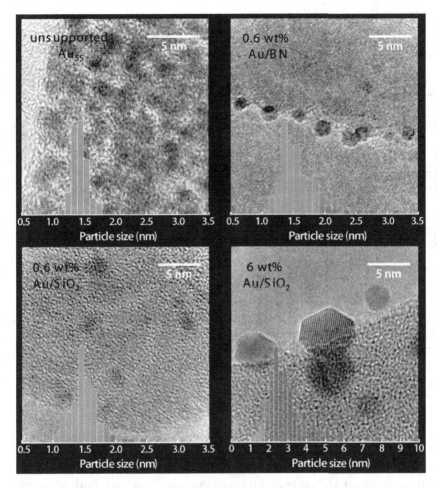

Fig. 4.3. TEM images showing the size distribution of Au nanoparticles synthesized on different supports with various loadings (courtesy of Prof. Owain Vaughan, University of Cambridge, UK).

selectivity of nanostructured catalysts. For example, photocatalytic properties of TiO_2 are improved by changing the shape of the film (Figure 4.2); on the other hand, controlled synthesis results in nanometer-sized Au particles with improved properties. The activity of Au–Pd catalyst depends on the nanoscale composition and particles with a Pd layer surrounding an Au core giving maximum productivity.

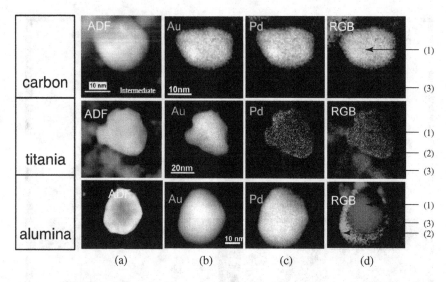

Fig. 4.4. STEM-ADF (annular dark-field) image of 2.5 wt% Au–2.5 wt% Pd/TiO$_2$ catalyst calcined at 400°C, showing (a) large-alloy particles, (b) Au–M$_2$ STEM-XEDS (X-ray energy-dispersive spectroscopy) maps, and (c) Pd L$_1$ STEM-EDS maps. Note that the Pd signal appears to originate from a larger area than that of the Au signal, as demonstrated in the STEM. In (d) grey-level intensities from Au, Pd, and Ti are converted into colors (blue (1), green (2), and red (3), respectively), and their composite images show their distribution in individual particles on carbon (*top row*), titania (*middle row*), and alumina (*bottom row*) (courtesy of Prof. Graham Hutchings, Cardiff University, UK).

4.3 Nanostructure Characterization Under Working Conditions

Ideally, catalysts should not change during a chemical reaction or should return to the same state between reaction cycles. However, both temperature and the nature of the gas/liquid environment often affect the morphology, structure, chemistry, and metal-support interactions during catalytic processes. Sometimes these changes are irreversible and restrict the long-term use of the catalyst. The reactive environment during synthesis also controls catalyst structure and morphology. Therefore, understanding the synthesis mechanisms that lead to high activity and selectivity of the catalysts as well as their functioning is a fundamental step forward to improve catalytic processes. The best way to understand the relationship of a nanostructured catalyst with its synthesis and its functioning is to follow the atomic-level changes occurring during synthesis and in a reactor, respectively.

Microscopy techniques are currently being used to study atomic- and near-atomic-level changes in two ways: (1) by characterizing the samples before and after the reaction, and (2) by making dynamic observations during the synthesis or operation of the catalyst. Regular STM, SEM, TEM, and STEM can be used for the first method, whereas special modifications are required for the dynamic imaging. Several examples are given below.

4.3.1 *Effect of environment on surface structure and reactivity*

Most catalytic reactions happen on surfaces; therefore, understanding the structure and chemistry of surfaces is an important part of the characterization. It has been shown that although CO does not adsorb on a clean Au{111} surface at 120 K and 5×10^{-8} mbar, it readily adsorbs under the same conditions if the surface has been pre-exposed to NO_2 (Zhang et al., 2005). Professor King's group at Cambridge University (as presented by Dr. Driver in the WTEC panel's visit; see Site Reports-Europe on the International Assessment of Research in Catalysis by Nanostructured Materials website, www.wtec.org/private/catalysis; password required, contact WTEC) has employed a combination of techniques such as reflection-absorption infrared spectroscopy (RAIRS), temperature-programmed desorption (TPD), LEED, and STM to understand the mechanism of surface activation as well as CO adsorption (Figure 4.5).

STM images (Figures 4.5(a) and 4.5(c)) show that a $(\sqrt{7} \times \sqrt{7})R19°$ surface reconstruction occurs due to interaction of the two co-adsorbed molecules (NO_2 + CO). Density functional theory calculations for various NO_2-to-CO ratios revealed that the structure model for a 3:1 ratio matches the images obtained (Figure 4.5(b)), but other ratios do not (e.g., the 2:1 model shown in Figure 4.5(d)). These results also show that co-adsorption of electronegative species on coinage metals can be advantageous, contrary to the belief that adsorption of one electronegative species may poison the catalyst for the adsorption of other electronegative species.

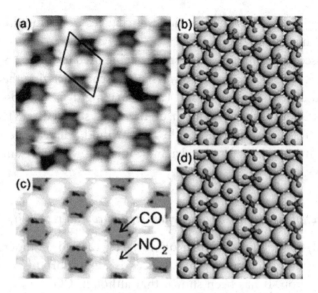

Fig. 4.5. (a) STM image (28×28 Å^2, gray-scale range 1.5 Å, sample bias + 0.10 V, current 1.0 nA) of an NO_2 + CO co-adsorbed layer, corresponding to the 3:1 structure shown in (b); (c) corresponding DFT Tersoff–Hamann simulation, (d) 2:1 model, for comparison (Zhang et al., 2005).

4.3.1.1 Sintering

Nanostructured catalyst particles are often subjected to reactive gaseous environments at elevated temperatures, both during synthesis and under operations. Catalyst particles can sinter under these conditions, which results in the reduction of the surface area and thereby activity. Therefore, in order to eliminate or reduce the sintering process, it is necessary to determine its mechanism and the effects of reaction conditions on the process. Generally, sintering can occur in two ways: (1) particles become mobile on the support surface and coalesce upon contacting another particle, and/or (2) smaller particles become unstable due to high surface energy (Gibbs–Thompson effect); atoms from the smaller particles diffuse to larger particles via either surface or vapor diffusion, depending upon the temperature and their melting point. The latter process is also known as Ostwald ripening, where smaller particles continuously reduce in size and finally disappear while larger particles continue to grow. Dynamic

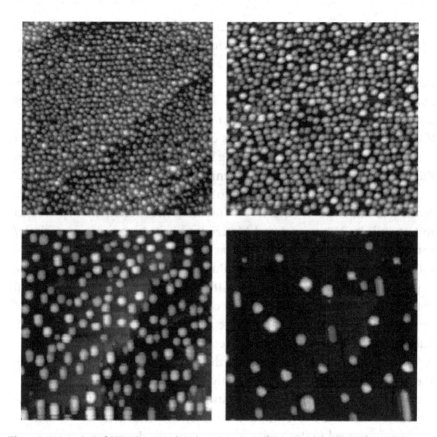

Fig. 4.6. A series of STM images showing sintering of an array of Pd nanoparticles after heating the layer deposited at room temperature to 973 K. The image size is 100 × 100 nm. The images were recorded after the temperature was stable over long periods of time. Temperatures of the images were (*top left*) annealed to 373 K; (*top right*) annealed to 673 K; (*bottom left*) annealed to 873 K; and (*bottom right*) annealed to 973 K. Note that sintering only becomes marked above ≈700 K (Bowker, 2006).

observations at various temperatures have been employed to understand the sintering behavior. Both of these processes have been observed and have been found to depend upon the particle morphology, support, and temperature.

For example, STM images of a Pd layer deposited by chemical vapor deposition (CVD) on a titania {110} surface were recorded during heating to follow the sintering mechanism, as illustrated in Figure 4.6 (Bowker,

2007). The images were recorded after the Pd layer was given ample time to achieve a stable temperature. It is clear from the images that de-wetting of the film resulted in the nucleation and growth of nanometer-size particles between 373 and 673 K, but the density of particles did not change considerably. However, particle size increased and density of particles decreased considerably as the temperature was increased from 673 K to 973 K. From these observations it is concluded that Pd particles do not sinter easily, and the fact that sintering happened only after 700 K indicates that it happened via the Ostwald ripening process rather than coalescence.

On the other hand, Dr. Kamino of Hitachi High-Technologies Corp., Japan (see Site Reports-Asia on the International Assessment of Research in Catalysis by Nanostructured Materials website, www.wtec.org/private/catalysis; password required, contact WTEC), showed time-resolved (30 frames per second) videos of Pd and Au particles moving on the support surface at 693 K in 1.1×10^{-2} Pa of air and coalescing upon coming in physical contact with each other. It appears that the mobility of particles on the support controls the coalescence process, while atomic-scale diffusion results in Ostwald ripening.

Liu et al. (2005) have shown that the sintering can proceed by either of the two mechanisms on the same catalyst support system, depending upon the morphology. Pd/Al_2O_3 is widely used as an industrial catalyst/support system for selective hydrogenation of acetylene. Reactivity of the catalyst has been observed to drop with time due to coke deposition on the catalyst surface. Heating at low temperatures (below 500°C) in steam is used to regenerate the catalyst. But the activity of the regenerated catalyst was always observed to be lower than that of fresh catalyst due to reduction in surface area. Direct observations of the sintering process were made using an environmental TEM (ETEM, Figure 4.1). *In situ* measurements showed that for fresh catalyst particles, significant sintering occurred only above 600°C by Ostwald ripening. On the other hand, sintering by coalescence was observed during the regeneration of used catalysts in steam at 500°C. As expected, used catalyst particles and support were covered with amorphous carbon that oxidized during heating in steam, and clean Pd particles became mobile on the support surface due to reduced metal-support interaction. Time-resolved (30 frames per second) observation showed that Pd particles move freely as the carbon coating burns out and coalesce when they come in contact with other particles. In fact, these regenerated particles were observed to catalyze the removal of carbon from the support surface, leaving channels for other particles to move freely and coalesce (Liu et al., 2005).

These observations show that the sintering process may be controlled by metal support interaction, morphology, temperature, and the natural gaseous/liquid environment. In order to maintain the activity and selectivity of the catalyst nanostructure, it is necessary to prevent an increase in particle size as well as any change in the structure of the exposed surfaces. Direct observations as described above can assist researchers in understanding the process and finding means to control particle growth.

4.3.1.2 Oxygen spillover

STM is the most suitable tool to obtain atomic-level images of surfaces, but the microscope tip can itself enhance the surface reaction process. Somorjai and colleagues (McIntyre et al., 1994) have shown enhanced activity for the hydrogenation reaction as the STM tip scanned the surface of Pt catalyst, due to hydrogen spillover in its vicinity. Bowker and colleagues presented *in situ* images of oxygen spillover for 4 nm Pd nanostructures on TiO_2 (110) surface (Figure 4.7) (Bennett et al., 1999).

Time-resolved images recorded at 673 K in an oxygen atmosphere show that the oxidation reaction started in the region surrounding the catalyst particle. Oxygen molecules adsorbed on Pd particles dissociate more readily than on an oxide surface to form atomic oxygen, which is more reactive. Oxygen atoms are then carried by the STM tip and deposited in the vicinity of catalyst particles (spillover), forming new layers of titania close to the catalyst particle — observed as a bright ring around the particle (Figure 4.7(c) that grew with time (Figures 4.7(d)–4.7(f)) — resulting in an $\approx$20-fold increase in reaction rate.

4.3.1.3 Synthesis of catalyst particles

Chapter 2 shows that there are various techniques currently employed for the production of nanostructured catalysts, and these synthesis methods control morphology, structure, and chemistry. Atomic-level variations in

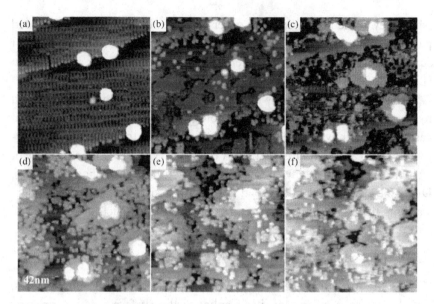

Fig. 4.7. A sequence of images showing the reaction of oxygen with Pd nanoparticles on TiO_2 (110) at 673 K. Image size is 42×42 nm; (a) is before reaction, and (b–f) is a sequence after exposure to a low-pressure stream of oxygen. The oxygen adsorbs at the Pd, and spillover to the adjacent titania then occurs (seen most clearly in (c)). The spillover oxygen grows new titania layers close to the particle much faster than elsewhere on the surface (Bowker, 2007).

the structure and composition can have a huge impact on the life and performance of the catalyst; therefore, synthesis conditions must be optimized to produce nanostructures with desirable activity and selectivity. It is quite common to employ electron and tunneling microscopy techniques to characterize the catalyst nanoparticles produced by various synthesis techniques and use the knowledge obtained to select the synthesis method. For example, impregnation of inorganic salt solution is often used to synthesize metal oxide (after calcinations) or metal (after reduction) on high-surface-area supports such as silica or alumina.

Professor de Jong's group at Utrecht University in The Netherlands (see Site Reports-Europe on the International Assessment of Research in Catalysis by Nanostructured Materials website, www.wtec.org/private/catalysis; password required, contact WTEC) has used high-angle annular dark-field images to determine the size and location of NiO on SBA-15 nanoporous support (Figure 4.8), using different calcination routes to

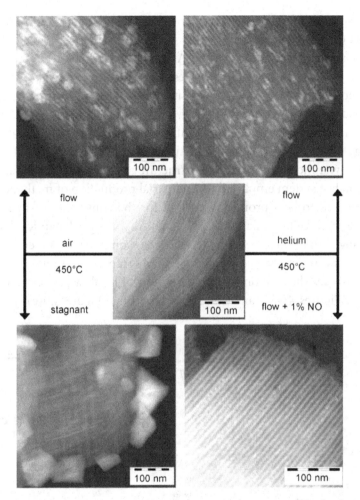

Fig. 4.8. HAADF images of NiO particles formed under different synthesis conditions showing that the best dispersion and smallest size of particles are obtained after calcinations in the flow using a mixture of He and NO (Friedrich et al., 2007).

form NiO from decomposition of an aqueous solution of $Ni(NO_3)_2(OH)_4$. They found that calcination in flowing, compared to stagnant, air at 450°C reduced the particle size from 20–100 nm to 10–35 nm; the number and size of particles formed were observed to reduce further to 11–26 nm by replacing air with He. The most interesting result was that a uniform distribution of 4 ± 1 nm-sized particles was obtained by adding 1% NO to the He flow during calcination. They also used electron tomography to show that most of the particles were located inside the pores of the SBA-15 support (Friedrich *et al.*, 2007).

Researchers at Haldor Topsøe A/S in Denmark (see Site Reports-Europe on the International Assessment of Research in Catalysis by Nanostructured Materials website, www.wtec.org/private/catalysis; password required, contact WTEC) have used *in situ* observations to understand the effect of environment on the surface structure and morphology of a catalyst during calcination. They have employed ETEM to observe dynamic evolution of nanostructure at the atomic scale. Cu/ZnO is the catalyst system employed for industrial production of methanol and is considered to be a promising catalyst for the conversion of hydrocarbons in fuel cell applications. The nature of the dynamic changes occurring with the change in the reducing gas composition under reaction conditions was determined using the ETEM (Hansen *et al.*, 2002).

Figure 4.9 shows atomic-scale evolution of catalyst particles in nonwetting (Figures 4.9(a) and 4.9(c)) and wetting (Figure 4.9(e)) catalyst

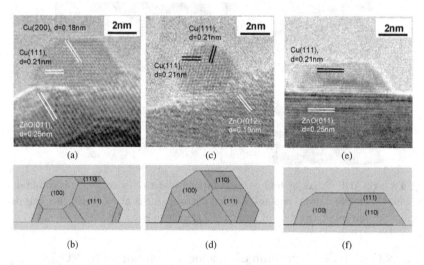

Fig. 4.9. *In situ* TEM images (a, c, and e) of a Cu/ZnO catalyst in various gas environments, together with the corresponding Wulff construction of the Cu nanocrystals (b, d, and f). The images shown in (a, c, and e) were recorded at 220°C in a pressure of, respectively, 1.5 mbar of H_2; in a gas mixture of H_2 and H_2O, ($H_2:H_2O = 3:1$; total pressure of 1.5 mbar); and in a gas mixture of 95% H_2 and 5% CO, at a total pressure of 5 mbar. The electron beam was parallel to the [011] zone axis of Cu (Hansen *et al.*, 2002).

morphology that exposed different surfaces as the reduction environment was changed by adding water or CO to hydrogen. The particle shape and surface structure was observed to change back to the original form as H_2O (less reducing) or CO (more reducing) was removed from the hydrogen flow, confirming the reversible nature of the restructuring phenomenon. Wulff constructions (Figures 4.9(b)–4.9(f)) show the availability of low index surfaces, such as 111, 110, and 100, and absence of high index surfaces. The surface free energy calculated from these images depends on the contact angle. For example, the surface energy of all low index surfaces was found to be lowest for particle morphology obtained in H_2/H_2O gas mixtures (Figure 4.9(b)).

4.3.1.4 Redox process

Ceria-based oxides are commonly used as supports for three-way catalysts used in automobile catalytic converters or as anode materials in solid-oxide fuel cells. In both applications, the ease with which CeO_2 gives up lattice oxygen (reduction) in an oxygen-lean environment and takes it back in an oxygen-rich environment (oxidation) — also known as oxygen storage capacity (OSC) — determines its activity. Recently, *in situ* observations have been made using ETEM to understand atomic-level structural and chemical changes during redox reactions responsible for CeO_2's high OSC (Wang *et al.*, 2008). A combination of high-resolution images, electron diffraction, and electron energy-loss spectra recorded between 400–800°C in 0.5–1.5 Torr of dry hydrogen gas has provided an unprecedented insight into the process.

For example, both an electron diffraction image and a high-resolution electron microscopy (HREM) image revealed the formation of a superstructure (Figure 4.10, center) during reduction. Electron energy-loss spectra recorded from the same catalyst particle confirmed Ce to be in +3 oxidation state as the Ce $M_{4,5}$ (white-line) intensity ratio reversed compared to the one recorded at 270°C (Figure 4.10, lower right and center). Moreover, the surface of the catalyst particle changed from rugged (Figure 4.10, upper right) to smooth (Figure 4.10, upper center) during reduction and did not reverse back upon reoxidation (Figure 4.10, upper left). The rugged facets expose the {111} polar surface, which is the most stable

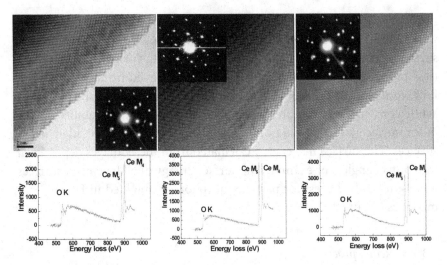

Fig. 4.10. HREM images (*top*), electron diffraction patterns (*inset*), and electron energy-loss spectra (*bottom*) of CeO_2 particles oriented along a 110 zone axis recorded at 270°C (*left*), 730°C (*center*), and 600°C (*right*) in 0.5 mbar of dry hydrogen. Note the appearance of the superstructure in the center image and the electron diffraction due to oxygen vacancy ordering upon reduction at high temperature that disappeared (right image) upon cooling as ceria oxidized back. Change in the ratio of Ce $M_{4,5}$ edges was used to determine the oxidation state of Ce (Wang *et al.*, 2008).

surface, and become unstable upon reduction. On the other hand, the neutral {110} surface is smooth and easy to reduce. Therefore, the ceria surface reconstructs to form stable {110} surfaces upon reduction that can be easily subjected to redox cycles.

4.3.1.5 *Catalyst poisoning: Synthesis of carbon nanotubes*

Transition metal catalysts are very suitable for a number of reaction processes in the petrochemical industry, such as hydrogenation, hydrocarbon hydrocracking, and Fischer–Tropsch synthesis. However, carbon generated during these processes deposits on the catalyst surfaces, making them inactive. Under selective conditions the carbon deposit can nucleate and grow to form carbon nanotubes (CNTs) that have a number of applications in bioscience and nanotechnology. Understanding the mechanism of coke deposition on transition metal particles can help researchers find ways to control the process. It has been proposed that carbon first dissolves in the metal and diffuses out after the lattice is supersaturated to form CNTs. However, the group at Haldor Topsøe has made direct observations of the growth of carbon nanofibers using ETEM. Time-resolved images recorded during decomposition of CH_4 at 536°C show that carbon deposited on the surface steps of Ni catalyst particles nucleates and grows to form graphitic carbon nanofibers (Helveg et al., 2004). Theoretical calculations performed by Prof. Nørskov's group at the Technical University of Denmark (see Site Reports-Europe on the International Assessment of Research in Catalysis by Nanostructured Materials website, www.wtec.org/private/catalysis; password required, contact WTEC) have confirmed that the diffusion of carbon on Ni surfaces is energetically favorable compared to dissolution of carbon in the Ni lattice (Abild-Pedersen et al., 2006). The Nørskov group also shows that surface steps play an important role for nucleation and growth. The WTEC panel found that this is an active area of research within the catalysis community in China but is not as active a focus area in Europe.

Similar dynamic observations are currently being applied for understanding the effects of temperature and gaseous environments on both nanostructured particles and supports. Combined with spectroscopy data, such observations can assist in understanding how to better control catalytic processes.

4.4 Future Trends

The WTEC panel found that characterization of nanostructured catalysts under working conditions is a priority for many groups in Asia, Europe, and the United States, and different groups are approaching it in different ways. Research groups in the UK (Cambridge and Cardiff) are most active in designing experiments using scanning probe microscopes (e.g., STM and AFM) to obtain atomic-level information about surface structures at a wide range of temperatures (liquid nitrogen −1000 K) in both gaseous and liquid environments. However, researchers must be careful about the effect of tip bias on the reaction process. Also, obtaining chemical information and changes below the surface layers is very difficult.

Electron microscopes (SEM, TEM, and STEM) operate best under high-vacuum conditions and are used to obtain structural and chemical information from the nanometer to the atomic scale. While following the effects of temperature on structure and chemistry of individual catalyst particles is quite simple, determining effects in gaseous environments requires extensive modification to the sample holder or to the microscope column and still these modifications can only achieve a fraction of the gas pressures used in a reactor. Moreover, researchers must be very careful about the effect of high-energy electrons on the sample and on the reaction process. Although most of the host institutes in China mentioned *in situ* observation under reactive conditions to be important, they did not yet have the facilities. Hitachi in Japan has developed holders and a modified microscope to observe a three-way-catalyst under a reactive environment. Both Haldor Topsøe and the Technical University of Denmark currently lead the world in applying *in situ* electron microscopy measurements to catalysis. In the United States, the effort is mostly at Arizona State University, but other institutions (e.g., Purdue University) and DOE National Labs (Brookhaven and Oak Ridge) are setting up facilities for *in situ* TEM observations under reactive environments.

It is clear from the above discussion that no single microscopy technique is perfect. This realization has led to the development of systems that combine a number of techniques to observe the samples under the same reaction conditions. This is often achieved by integrating sample transfer chambers between two complementary techniques that require different operating conditions. For example, after imaging the effects of a reactive environment using STM, samples can be transferred to an X-ray photoelectron spectroscope (XPS) to obtain spectroscopy information; this is being done at ETH in Zurich, Switzerland, and the Technical University of Munich in Germany (see Site Reports-Europe on the International Assessment of Research in Catalysis by Nanostructured Materials website, www.wtec.org/private/catalysis; password required, contact WTEC). Researchers at the Fritz Haber Institute in Berlin (see Site Reports-Europe on the International Assessment of Research in Catalysis by Nanostructured Materials website, www.wtec.org/private/catalysis; password required, contact WTEC), are currently developing an ultrahigh-resolution spectromicroscope (known as SMART), a high-field W-band electron paramagnetic resonance spectrometer to investigate surfaces in ultrahigh vacuum, and a new low-temperature STM. The SMART utilizes both chromatic and spherical aberration corrections to overcome the resolution limitations of currently available photoelectron emission microscopes to achieve 2 nm spatial resolution and 100 meV energy resolution. They also plan to use this instrument for near-edge X-ray absorption fine-structure, XPS, and Auger spectroscopy, as well as for diffraction methods such as microspot-LEED and photoelectron diffraction.

At Cardiff University, a hot filament deposition chamber is connected to an STM, EPR (electron paramagnetic resonance), LEED, and a preparation chamber. The same sample can be transferred from one system to the other without being exposed to air, in order to obtain imaging and various spectroscopic information from the same regions of interest.

4.5 Summary

Electron and tunneling microscopy are used for routine characterization of nanostructured catalysts to correlate synthesis to properties and functions. These techniques are quite effective in obtaining nanoscale structural and chemical information during various levels of synthesis and operation. Although there is a widespread consensus on the need to observe catalysts under reactive environments, very few groups have made investments to purchase or develop the required instrumentation. Environmental SEM (ESEM) and STM are more readily available than environmental TEM (ETEM), most probably because of TEM/ETEM's high cost and operational difficulty. Also there is clearly a need to improve both temporal and spatial resolution to be able to observe atomic-level changes in the reaction time frame.

References

Abild-Pedersen, F., Nørskov, J. K., Helveg, S., Sehested, J., and Rostrup-Nielsen, J. R. (2006). Mechanisms for catalytic carbon nanofiber growth studied by ab initio density functional theory calculations, *Phys. Rev. B*, 73, p. 115419.

Bennett, R., Stone, P., and Bowker, M. (1999). Pd nanoparticle enhanced re-oxidation of non-stoichiometric TiO_2: STM imaging of spillover and a new form of SMSI, *Catal. Lett.*, 59, pp. 99–105.

Bowker, M. (2007). Resolving catalytic phenomena with scanning tunnelling microscopy, *Phys. Chem. Chem. Phys.*, 9, pp. 3514–3521.

Enache, D. I., Edwards, J. K., Landon, P., Solsona-Espriu, B., Carley, A. F., Herzing, A. A., Watanabe, M., Kiely, C. J., Knight, D. W., and Hutchings, G. J. (2006). Solvent-free oxidation of primary alcohols to aldehydes using Au-Pd/TiO_2 catalysts, *Science*, 311, pp. 362–365.

Friedrich, H., Sietsma, J. R. A., de Jongh, P. E., Verkleij, A. J., and de Jong, K. P. (2007). Measuring location, size, distribution, and loading of NiO crystallites in individual SBA-15 pores by electron tomography, *J. Am. Chem. Soc.*, 129, pp. 10249–10254.

Gai, P. L. (1999). Environmental high resolution electron microscopy of gas-catalyst reactions, *Top. Catal.*, 8, pp. 97–113.

Hansen, P. L., Wagner, J. B., Helveg, S., Rostrup-Nielsen, J. R., Calusen, B. S., and Topsøe, H. (2002). Atom-resolved imaging of dynamic shape changes in supported copper nanocrystals, *Science*, 295, pp. 2053–2055.

Helveg, S., Lopez-Cartes, C., Sehested, J., Hansen, P. L., Clausen, B. S., Rostrup-Nielsen, J. R., Abild-Pedersen, F., and Nørskov, J. (2004). Atomic-scale imaging of carbon nanofibre growth, *Nature*, 427, p. 426.

Hughes, M. D., Xu, Y. J., Jenkins, P., McMorn, P., Landon, P., Enache, D. I., Carley, A. F., Attard, G. A., Hutchings, G. J., King, F., Stitt, E. H., Johnston, P., Griffin, K., and Kiely, C. J. (2005). Tunable gold catalysts for selective hydrocarbon oxidation under mild conditions, *Nature*, 437(7062), pp. 1132–1135.

Hutchings, G. J. (2007). A golden future for green chemistry, *Catal. Today*, 122, pp. 196–200.

Liu, R.-J., Crozier, P. A., Smith, C. M., Hucul, D. A., Blackson, J., and Salaita, G. (2005). Metal sintering mechanisms and regeneration of palladium/alumina hydrogenation catalyst, *App. Catal.*, A282, p. 111.

McIntyre, B. J., Salmeron, M., and Somorjai, G. A. (1994). Nanocatalysis by the tip of a scanning tunneling microscope operating inside a reactor cell, *Science*, 265(5177), pp. 1415–1418.

Perathoner, S., Passalacqua, R., Centi, G., Su, D. S., and Weinberg, G. (2007). Photoactive titania nanostructured thin films: Synthesis and characteristics of ordered helical nanocoil array, *Catal. Today*, 122, pp. 2–13.

Sharma, R. (2005). An environmental transmission electron microscope for *in situ* synthesis and characterization of nanomaterials, *J. Mater. Res.*, 20(7), pp. 1695–1707.

Wang, R. G., Crozier, P. A., Sharma, R., and Adams, J. B. (2008). Measuring the redox activity of individual nanoparticles in cerium-based oxides, *Nano Lett.*, 8(3), 962–967.

Zhang, T., Liu, Z.-P., Driver, S. M., Pratt, S. J., Jenkins, S. J., and King, D. A. (2005). Stabilizing CO on Au with NO_2: Electronegative species as promoters on coinage metals, *Phys. Rev. Lett.*, 95, p. 266102.

5

Theory and Simulation in Catalysis

Matthew Neurock

5.1 Introduction

The past two decades have witnessed tremendous advances in the development and application of theory and computational catalysis to catalysis. Computational catalysis has reached the stage where it has become an invaluable complement to experimental efforts aimed at understanding and controlling the reactivity of catalytic materials. The increase in computer processor speeds, the development of parallel architectures, and advances in theoretical methods have enabled the simulation of much more complex heterogeneous catalytic materials and processes. As a result, theory and simulation have lead to unprecedented advances in elucidating catalytic reaction mechanisms, establishing the influence of realistic reaction environments, and aiding in the design of new materials.

Despite this progress, the complexity of many catalytic materials along with the complexity of the reaction environments in which they operate present significant challenges for theory and simulation. Figure 5.1 presents just a few examples of current catalytic systems in order to highlight

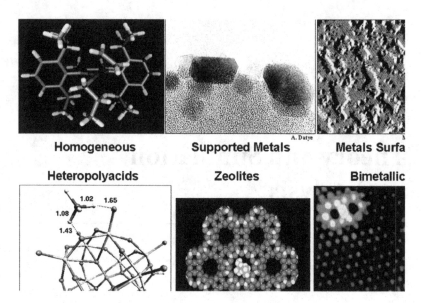

Fig. 5.1. Structure and complexity of different catalytic materials.

some of the differences in the types of materials and the challenges they present for theory and simulation. This includes homogeneous and supported homogeneous catalysts with complex ligand spheres; metal particles of different sizes, morphologies and compositions anchored to different supports; well-defined nano-oxide clusters and porous materials such as polyoxometallates and zeolites; and ill-defined metal oxides, sulfides, carbides, and nitrides, just to name a few. In addition to the material structures, the reaction media and the local environments in which the catalytic transformations are carried out is becoming increasingly more complex, involving the presence of solution, mixed solvents, ionic liquids, and complex metals.

The ability to design such complex materials will undoubtedly require a fundamental understanding of the active site, the elementary transformations that occur at the site, and the influence of the reaction environment under working conditions. Advances in the design of such systems and their industrial applications will therefore require traversing a very diverse spectrum of time and length scales, ranging

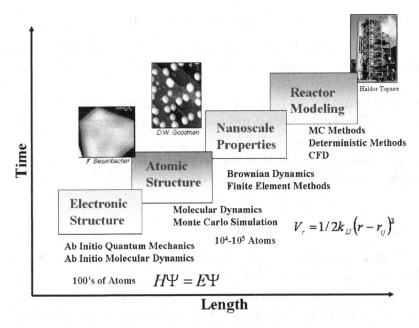

Fig. 5.2. Hierarchy of time and length scales in heterogeneous catalysis, and associated modeling methods.

from the electronic structure that controls the chemistry at the active site, to the detailed atomic structure of supported particles, on up to the performance of the entire catalytic process. The methods that are currently used to model particular features such as the atomic structure of a material or temperature and composition profiles in a reactor, currently exist as shown in Figure 5.2. However, the ability to traverse across different time and length scales, which is important in accessing how the changes in the atomic structure of the catalyst ultimately influence catalytic process performance, presents a number of difficult challenges. While the seamless integration across the full spectrum of time and length scales presents a tremendous challenge that will not be solved anytime in the near future, we are beginning to see the combining of methods to cross time- and length-scale boundaries, as will be discussed in this chapter.

5.2 Computational Catalysis: Where Are We Today?

The results from the WTEC panel's assessment clearly reveal that theory and simulation have made important advances and are now able to:

- Predict structure and properties of model catalytic systems
- Complement and resolve spectroscopic measurements by simulating the resulting spectra
- Elucidate catalytic reaction pathways and mechanisms
- Begin to guide the search for new catalytic materials.

5.3 Methods and Their Applications

Computational catalysis is typically defined by the methods that link the detailed electronic structure to catalytic performance (the first 2–3 boxes in Figure 5.2). This involves the determination of structure and chemical properties, intrinsic catalytic reactivity, and the simulation of catalytic kinetics, which are best modeled by quantum mechanical methods, atomic scale simulations, and kinetic simulations. A very short review of each of these areas and their application is presented below. More detailed discussions on these methods are presented elsewhere (e.g., van Santen and Neurock, 2006).

5.3.1 *Electronic structure methods*

The ability to simulate reactivity ultimately requires the solution of Schrödinger's equation as it involves making and breaking chemical bonds and thus the ability to follow changes in the electronic structure. Reactivity can only be resolved then by methods that follow the changes in the forces on the electrons and the nuclei. There are a wide range of electronic structure methods that can be classified as (1) semi-empirical, (2) *ab initio* wave function, and (3) density functional theory (DFT) methods (Head-Gordon, 1996; Leach, 1996; Jensen, 1999). All of these methods start with the time-independent solution to the Schrödinger's equation and make a series of assumptions in order to solve the force balance that results on the N-particle (proton and electron) system. Ultimately, our inability to directly solve the multicenter integrals that result from the electron-electron interactions lead to the three different approaches presented. Semi-empirical methods avoid the solution of the multicenter integrals by fitting the results to experimental data. The success for transition-metal-based systems, however, has been very limited. While early studies in catalysis used these methods to provide qualitative understanding, they have for the most part been replaced by *ab initio* wave function or density functional theory studies.

Ab initio wave function methods involve the solution of the structure and the energy of the N-electron wave function by making a sequence of approximations that allows one to decouple the N-electron system into one that involves the solution of N single electron systems. These assumptions lead to the simplest solution strategy, which is known as the Hartree–Fock (HF) solution. The HF solution avoids the direct electron-electron interactions, solving them instead via a mean-field approach (Head-Gordon, 1996; Leach, 1996; Jensen, 1999). This is a gross oversimplification that misses out on the correlated electron motion that results from the direct interactions between electrons. These interactions can be accounted for but require much more rigorous calculations.

The accuracy of *ab initio* wave function methods can be systematically improved by moving to more complete descriptions of the configurational interactions (CI) and basis functions used to describe the wave function. An infinite accounting of the configurational interactions and an infinite basis set expansion would lead to the exact energy of the N-particle problem. As the CI and basis function limits are extended, the size of the system that can be solved becomes significantly smaller (Head-Gordon, 1996; Jensen, 1999). While CI calculations can be carried out on cluster models for catalytic systems, they are limited to clusters that are less than 50 heavy atoms. While these methods have been useful for homogeneous catalytic systems, there have been very limited applications to more complex heterogeneous systems, as they are limited by system size.

Quantum Monte Carlo (QMC) methods offer a second approach to accurately model electron exchange and configuration interactions. While QMC methods are formally less costly than *ab initio* CI methods, they have their own set of drawbacks in terms of finite size effects that require large unit cells, the lack of analytic second derivatives and force predictions, and significant challenges in modeling transition metals (Carter, 2008).

The development of density functional theory has had a tremendous impact on computational heterogeneous catalysis as it has allowed for the solution of systems that offer more realistic models of the active site and the catalyst surface. In the original formulation, Hohenberg and Kohn (Hohenberg and Kohn, 1964; Kohn and Sham, 1965) showed that the ground state energy for a system is uniquely described by a functional of its electron density. Kohn and Sham later extended this to practice by showing how the energy could be divided into kinetic energy, attractive nuclear-electron interactions, and an exchange-correlation term, all of which could be described as a functional of the density. While this is an exact approach to solve for the energy of the system as a functional of the density, there is no formal way in which to determine the exchange correlation functional.

Current acceptable solutions involve calculating gradient-corrections to the local approximation for the correlation and exchange energies (Becke, 1986; Perdew *et al.*, 1992). Despite these corrections, the accuracy of DFT is still limited, and thus this is a very active area of research in theoretical chemistry. Since the electron density is the fundamental entity, DFT tends to scale as N^3 where N is the number of electrons. In comparison, *ab initio* many-body perturbation methods and higher-level CI calculations tend to scale from N^5–N^7. Density functional theory thus allows for the simulation of systems that contain up to a few hundred atoms. This proves to be useful in determining the adsorption energies and reactivity on well-defined single-crystal surfaces as well as within zeolites. In terms of absolute accuracy, density functional theory is typically within 20 kJ/mol (van Santen and Neurock, 2006). There are, however, known outliers. In addition, DFT fails to appropriately account for weak dispersion forces. For systematic trends, DFT is even more accurate and in many cases can be system-dependent. Coupled-cluster theory can be used to provide accuracy to within a 2 kJ/mol but is typically limited to systems of the size of 10 atoms (Head-Gordon, 1996; Leach, 1996; Jensen, 1999; van Santen and Neurock, 2006). While method accuracy is certainly important, it must also be balanced with model accuracy that refers to the size of the system used to model the active site. DFT has thus provided an important impact in the field, because it tends to balance both method and model accuracies.

The advances in both wave function methods and density functional theory now allow us to simulate both the electronic and the geometric structure for a specific system with a reliable degree of accuracy. They can also be used to elucidate transition states and thus determine activation barriers as well as rate constants by invoking transition state theory or variational transition state theory. The advances in density functional theory as well as the advances in computation have now made it possible to simulate complex networks of elementary steps over model substrates in order to determine actual catalytic performance, such as activity and selectivity, and in addition, these advances provide the ability to resolve reaction mechanisms.

Ab initio methods in principle can be used to simulate a wide range of structural, electronic, or energetic properties of the system. This has led to tremendous advances in simulating the spectroscopic interrogation of the state of the surface or the adsorbed intermediates on the surface. For example, the calculation of second derivatives of energy with respect to changes in the atomic positions can be used to calculate the vibrational frequencies of an adsorbed intermediate, which can then be directly compared with those from infrared or Raman spectroscopy. DFT is typically within about 5% of the actual spectra (van Santen and Neurock, 2006). Similarly, theory can be used to simulate nuclear magnetic resonance shifts and optical spectra such as ultraviolet visible (UV) shifts and nuclear magnetic shifts. The accuracy for DFT predictions of these properties is typically less. Band gaps, for example, are only good to within about 0.5 eV. Higher-level methods and new functionals, however, have resulted in significantly more accurate descriptions for particular systems (Staroverov *et al.*, 2004; Hafner, 2008).

Embedding methods have been developed in order to expand the model size without sacrificing the accuracy of the electronic structure and reactivity at the active site that carries out the chemistry. Embedding involves linking an active site region, which is described by a higher level *ab initio* method such as density functional theory, to an outer shell, which is described by a lower-level method such as HF or molecular mechanics (Sauer, 1994; Whitten and Yang, 1996; Govind *et al.*, 1999; Froese and

Morokuma, 1999; Vreven and Morokuma, 2000). This is typically done by defining a region of overlap where the two descriptions meet. While these methods provide the ability to simulate much larger systems, there are still issues with appropriately establishing how the inner reactive zone is linked to the outer region. This approach has been quite useful in simulating zeolites where the outer Si-O framework can appropriately be handled by lower-level calculations (Sauer, 1994).

5.3.2 *Atomic and molecular simulations*

While the simulation of reactivity requires an accurate accounting of the electronic structure, there are many important questions that are focused on structure or thermodynamics rather than on reactivity. As such, the electron density is not very important and the fundamental entity then becomes the atom. Atomic-scale simulations track the interatomic interactions that govern both structure and dynamics and ignore the electronic structure. The interactions are modeled by force fields, which are determined *a priori* from experimental regression or through *ab initio* calculations. The force fields are based on atomic (intra- and inter-molecular) interactions to the potential energy that result from changes in the bond length, bond angle, torsion angle from their standard positions, as well as Coulombic and van der Waals intermolecular forces (Leach, 1996; Cummings, 2002).

Energy minimization algorithms are perhaps the most basic simulation methods employed in catalysis. They are typically used to optimize complex material structures. Simulated annealing strategies are often employed to help isolate the lowest energy states. Energy minimization schemes are routinely used to help determine the structure of metal and metal oxide surfaces and particles as well as complex and porous oxides. While these simulations are easy to carry out and offer very useful information, they require reliable interatomic potentials for the materials of interest.

The thermodynamic properties for an N-particle (atoms or molecules) system can be determined from statistical mechanics. Monte Carlo simulation methods are typically used in order to integrate the configurational integral and solve for various thermodynamic properties. Various methods have been used in the literature to simulate different properties. The properties of interest typically dictate which methods are used (Leach, 1996; Bell et al., 1997; Cummings, 2002).

The three most widely adopted methods in catalysis are the Canonical, Grand Canonical and Gibbs Ensemble Monte Carlo simulations which are used to determine the pressure, sorption properties, and phase behavior for different catalytic systems (Leach, 1996; Bell et al., 1997; Cummings, 2002). Canonical simulations hold the number of molecules, volume, temperature, and unit cell constant and are typically used to minimize the Helmholtz energy of a system and establish the resulting pressure. Grand Canonical Ensemble Monte Carlo (GCMC) simulations hold the chemical potential, the volume, and the temperature constant in order to determine the optimal number of particles. GCMC simulations are typically used to simulate equilibrium sorption in microporous systems such as zeolites (Bell et al., 1997). Gibbs Ensemble Monte Carlo simulations act to minimize the Gibbs free energy and have also been used to simulate sorption behavior as well as phase equilibria.

5.3.3 *Dynamics*

The simulation of dynamics requires models that can follow temporal changes in structure. Classical systems are solved by simply integrating Newton's equation of motion to determine the time-dependent behavior of the atoms or molecules (Allen and Tildesley, 1987; Frenkel and Smit, 1996; Leach, 1996). These simulations are based on forces that act upon the atoms at any given time and thus require accurate force fields. The integration of the forces that act upon each atom within the system provides the time-dependent changes in position and the velocities in the systems. Molecular dynamics (MD) simulations have been used to simulate a range of dynamic properties for catalysis, including diffusion, transport, temporal changes in the catalyst surface structure, and the structural changes in an external fluid media with time (Bell et al., 1997).

Reaction systems require the ability to follow the dynamics of the atomic nuclei as well as the electrons. There are currently three different approaches to simulate the dynamics of reaction systems: (1) quantum dynamics, (2) *ab initio* molecular dynamics, and (3) Car–Parrinello molecular dynamics (van Santen and Neurock, 2006). Quantum dynamics fully treats both the dynamics of the electrons as well as the nuclei. These are very expensive simulations and have been used predominantly to follow the dynamics of in the collisions of very small molecules such as hydrogen, nitrogen, and methane on ideal surfaces.

Ab initio molecular dynamics methods are far less expensive, because they invoke the Born–Oppenheimer approximation in order to decouple the movement of the electrons and the nuclei. *Ab initio* methods are then used to calculate the forces on each of the atoms at a particular instant in time. The forces are used to solve Newton's equation of motion for very small time steps which provides the new positions of the nuclei. *Ab initio* methods are subsequently used to calculate the energies as well as the forces at this new position and time. This constant alternation between *ab initio* calculations and classic dynamics simulations is continued in order to map out the time dependent properties of the reaction system.

The final approach, developed by Car and Parrinello (1985), treats the electrons as "particle-like" and assigns fictitious masses to them. This allows the electrons to be propagated together in time with the nuclei, which significantly speeds up the calculations since quantum mechanical (QM) calculations are not required after every step. The latter two approaches have been used to simulate both homogeneous as well as heterogeneous catalytic systems that occur in the solution phase.

5.3.4 *Kinetics*

The ability to move from detailed electronic structure to catalytic kinetics requires some form of coarse graining in order to traverse the disparate time and length scales. The classic approach to simulating kinetics involves the solutions to a set of differential equations that describe the formation and destruction of all of the reaction intermediates that result from the controlling reaction paths. Dumesic (Dumesic *et al.*, 1993) pioneered the microkinetic approach, which involves modeling the elementary paths that comprise the mechanism. The activation energies from detailed *ab initio* calculations can be used together with transition state theory to provide the input required to solve the resulting differential equations and follow the kinetic behavior of the system.

A second approach to modeling kinetics takes advantage of stochastic kinetic approaches that maintain the detailed structure and positions of atoms in the surface and the adsorbed molecules. This avoids the loss of atomic structural information concerning the surface that is inherent in deterministic models due to the early averaging that is required (van Santen and Neurock, 2006). As such, one can explicitly track the atomic surface structure and individual molecular transformations. The same set of elementary surface processes can now be simulated together with the atomic surface structure. Monte Carlo techniques are typically used to solve these stochastic systems. A wide range of different catalytic kinetic systems have been modeling using kinetic Monte Carlo simulations methods, including temperature programmed desorption, temperature programmed reaction, transient kinetics, and steady state kinetics (van Santen and Neurock, 2006).

5.4 Snapshot of the Efforts in Europe and Asia

While there are outstanding efforts in the development and application of theory and simulation to catalysis going on throughout the world, the WTEC study focused on Europe and Asia. A summary of the sites that the WTEC panelists visited as well as other sites with strong programs in computational catalysis is given in Tables 5.1 and 5.2. The tables do not capture all of the scientists working in computational catalysis but instead provide a snapshot of what is happening in both regions. A detailed analysis of the findings indicates that there are characteristic differences among Europe, Asia, and the United States.

5.4.1 *Europe*

Computational catalysis has a long history in Europe. There is a core group of well-established scientists who have been instrumental in the success of theoretical methods and their application to catalysis over metals, metal oxides, metal sulfides, and zeolites. Most of the solid-state electronic structure codes that are currently used throughout the world to

Table 5.1. Important European sites and activities in computational catalysis, theory, and simulation.

Country	Site and Activity
Austria	Juergen Hafner, Georg Kress, University of Vienna: Structure, electronic chemical properties of metals, metal oxides, metal sulfides and zeolites; *ab initio* method development (VASP)
Denmark	*Jens Nørskov, Thomas Bligaard, Jens Rossmeisi, Technical University of Denmark: Catalytic reactivity of metals, metal alloys, metal oxides, metal sulfides, and zeolites for steam reforming; oxidation; direct NO decomposition; PEM and solid oxide fuel cells; ammonia synthesis and decomposition; *ab initio* method development (DECAPO)
	Bjork Hammer, University of Aarhus: Surface chemistry and catalytic reactivity of metals and metal alloys
France	Philippe Sautet, Francoise Delbecq, David Loffreda, Univ. of Lyon; Ecole Normale Supérieure of Lyon: Surface catalyis on metals, alloys, supported organo-metallics, aluminum oxide surfaces, integration of theory and spectroscopy
	*Pascal Raybaud and Hervé Toulhoat, IFP: Molecular modeling, *ab initio* thermochemistry of catalytic surface, catalytic reactivity, transition metal sulfides for hydrodesulfurization, aluminum and titanium oxides for catalytic supports, active phase-support interaction, hydrogenation
	Nicolas Bats, IFP: Guided synthesis and discovery of microporous materials
	(*) Christian Minot, Université Pierre et Marie Curie: *Ab initio* applications to metal oxides and metal oxide catalyzed reactions
	Odile Eisenstein, Jean-Sébastien Filhol, Marie-Liesse Doublet, Frédéric Lemoigno, Univ. of Montpellier: Supported organometallics, electrocatalytic and photochemical systems
	Jean-François Paul, University of Science and Technology of Lille: Hydrodesulfurization catalysis
Germany	Joachim Sauer, Humboldt University: Oxidation and acid catalysis over metal oxides, zeolites, method development
	*Matthias Scheffler (and Karsten Reuter), Fritz Haber Institute: Oxidation catalysis over metal oxides, surface kinetics, *ab initio* and statistical mechanical developments
	*Notker Roesch, Technical University of Munich: Catalysis over metal particles, steam reforming, metal particle/oxide support interfaces, DFT developments
	*Klaus Hermann, Fritz Haber Institute: Surface crystallography, structure/reactivity/spectro-scopy of metal oxides, theory development (StoBe)

(*Continued*)

Table 5.1. (*Continued*)

Country	Site and Activity
	Frerich Keil, Technical University of Hamburg: N_2O, and adsorption and diffusion simulations, in zeolites
	Axel Gross, University of Ulm: Chemisorption and reactivity on metals
Iceland	Hannes Jonsson, University of Iceland: Surface reactivity of metals; *ab initio* methods development
Italy	Gianfranco Pacchioni, University of Milan: Structure and reactivity of metal oxides, metals and supported metals, role of defects
	*Roberto Dovesi, Roberto Orlando, Carla Roetti, Cesare Pisani, V. Saunders, Piero Ugliengo, University of Turin: Structure and reactivity of metal oxides and zeolites, *ab initio* method development (crystal)
Spain	Francesc Illas, University of Barcelona: Catalysis over transition metal and alloy surfaces; metal-metal oxide interfaces, electrochemical effects
	*Avelino Corma, University of Valencia: Combined theory and experiment for zeolites and transition metals
	Javier Santz, University of Seville: Metal and metal oxide surfaces and interfaces
Switzerland	*Alfons Baiker, ETH: Theory and spectroscopy to understand the chiral modifiers to metal surfaces
Netherlands	*Rutger van Santen (and Tonek Jansen), Eindhoven University of Technology: Zeolite catalysis, catalysis over metals for a wide range of reactions, *ab initio* applications, and development of molecular simulations, theory, and experiment
	Mark Koper, Leiden University: Electrocatalysis theory and experiment
	Berend Smit, University of Amsterdam (now at Berkeley): Adsorption, diffusion, and reactions in porous materials
	Geert-Jan Kroes, Leiden University: Quantum dynamics of dissociative adsorption on metals
UK	Richard Catlow, University College London: Structure and reactivity of complex micro/mesoporous catalytic materials, zeolites; molecular simulation method development
	Peijun Hu, Queens University, Belfast: Structure and reactivity of metal surfaces for water gas shift, Au catalysis, Fischer–Tropsch synthesis
	*David King, Steven Jenkins, University of Cambridge: Reactivity of metal surfaces; combination of theory and experiment
	*David Willock, Cardiff University: Metals and zeolites

* Sites visited by WTEC panelists during the study
(*) Dr. Minot was not present; panelists did not see his lab.

Table 5.2. Important Asian sites and activities in computational catalysis, theory, and simulation.

Country	Site and Activity
China	Guichang Wang, Nankai University: Reactions over metal surfaces
	W. K. Chen, Fuzhou University: Methanol and N_2O decomposition on metal and metal oxide
	Zhi-Pan Liu, Fudan University: CO oxidation on metals, Fischer–Tropsch synthesis
	Zhaoxu Chen, Nanjing University: Water dissociation on metals
	Daiqian Xie, Nanjing University: Methane dissociation on metals
	*Haijun Jiao, Chinese Academy of Science, Taiyuan: Fischer–Tropsch synthesis
	Q. Jiang, Jilin University: Ethylene epoxidation on Ag(111)
Japan	Akira Miyamoto, Tohoku University: Development of novel *ab initio*-based molecular dynamics methods, simulation of NOx reduction, photocatalysis, oxidation, zeolites, combinatorial catalysis
	Hisayoshi Kobayashi, Kyoto Institute of Technology: Photocatalysis, splitting of water, doping metal oxides, oxidation over metals
	*AIST: Photocatalysis
Korea	*Jae Sung Lee, Pohang University: DFT calculations of band gaps, band edges, and optical absorbance for photocatalysis

*Sites visited by WTEC panelists during the study

model heterogeneous catalytic systems were developed in Europe by leading theoretical physics and chemistry groups. Those include the following:

- *CASTEP (Cambridge Serial Total Energy Package)*: Michael Payne and colleagues, Cambridge University (Segall *et al.*, 2002), www.castep.org Commercial code offered by Accelrys, Inc.
- *VASP (Vienna Ab Initio Simulation Package)*: Juergen Hafner, Georg Kresse, and colleagues, University of Vienna (Kresse and Hafner, 1993; Kresse and Furthmüller, 1996a; 1996b), cms.mpi.univie.ac.at/vasp
- *DACAPO*: Jens Nørskov and colleagues, Danish Technical University, wiki.fysik.dtu.dk/dacapo
- *CRYSTAL*: C. Pisani, R. Dovesi, C. Roetti, and colleagues, University of Turin (Pisani and Dovesi, 1980; Saunders, 1984; Pisani *et al.*, 1988), www.crystal.unito.it

- *ADF BAND (Amsterdam Density Functional Theory)*: Professor Evert Jan Baerends and colleagues, Free University of Amsterdam (Philipsen *et al.*, 1997; Velde *et al.*, 2001), www.scm.com.

Commercial code offered by SCM, Inc.

- *DMOL3*: Professor Bernard Delley and colleagues, Paul Scherrer Institute (Delley, 2000a; 2000b), accelrys.com/products/materials-studio/modules/dmol3.html.

Commercial code offered by Accelrys, Inc.

In addition to the development in Europe of solid-state methods, there have also been important developments in embedding methods and atomistic simulations that allow for the simulation of much larger and more realistic environments for both zeolites and metal oxides.

- *QMPOT*: Joachem Sauer and colleagues, Humboldt University (Eichler *et al.*, 1997; Sauer and Sierka, 2000)
- *GULP*: Richard Catlow and Julian Gale, University College London (Gale, 1997; Gale and Rohl, 2003).

Commercial code offered by Accelrys, Inc.

These codes are used throughout the world, and many have been adopted into commercial codes as well.

Europe has also led much of the efforts in the theory of heterogeneous catalysis, with strong efforts by well-established scientists who have been active in the field for over 30 years. Much of the current theory that governs metal, metal oxide, metal sulfide, and zeolite catalysis has been the result of leading efforts in Europe. A list of some of the established groups in each area is given below. An important point worth noting is that essentially all of the efforts in Europe reside in the basic sciences, i.e., Chemistry and Physics.

- *Metals*: Jens Nørskov (Denmark), Matthias Scheffler (Germany), Notker Rösch (Germany), Francesca Illas (Spain), Philippe Sautet (France)
- *Oxides*: Gianfranco Pacchioni (Italy), Matthias Scheffler (Germany), Klaus Herrmann (Germany), Joachim Sauer (Germany), Christian

Minot (France), Pascal Raybaud and Hervé Toulhoat (France), Malgorzata Witco and Jerzy Haber (Poland), Lars Peterson (Sweden)
- *Sulfides*: Jens Nørskov (Denmark), Pascal Raybaud and Herve Toulhoat (France).

In addition to these established programs, there is an increasing number of younger faculty members in Europe who have strong efforts in ancillary electronic structure developments and their application to more complex materials and their environments. Their activities are also listed in Table 5.1.

5.4.2 *Asia*

There was very little discussion of theory and simulation at the sites that the WTEC panelists visited in Japan, China, and Korea. The panel findings, as well as a few of the well-established sites known for strong efforts in computational catalysis in Japan, are highlighted in Table 5.2.

The theoretical and simulation efforts for catalysis in Japan are strong but tend to be concentrated in just few laboratories. This is a characteristic of Japanese research programs in general, in which there are just a few groups but very strong concentrated programs. In particular, Professor Akira Miyamoto and Professor Hisayoshi Kobayashi are internationally known for their important contributions to computational catalysis.

Professor Miyamoto has led the development of *ab initio* molecular dynamics codes, novel *ab initio* tight-binding molecular dynamics methods, and combinatorial informatics algorithms. The work of his group covers a broad range of different catalytic materials, including rare earth oxides, titanosilicates, metal oxides and metals, as well as a range of different applications, including NO_x reduction, decomposition of volatile chemicals, methanol synthesis, Fischer–Tropsch synthesis, photocatalysis, and Ziegler–Natta polymerization catalysis. The Miyamoto group is comprised of over 80 researchers; these include 9 visiting professors, 3 visiting assistant professors, 6 associate professors, 2 assistant professors, 3 research fellows, 3 cooperative researchers, and 53 students or technical staff. This is perhaps the largest effort in theory in the world, but not all of the researchers are focused on catalysis.

The work in the Professor Kobayashi group has a strong focused effort on modeling photocatalysis and electrocatalysis. The work is closely coupled with strong experimental efforts at different institutions in Japan. There are also strong efforts at AIST, which are focused on modeling photocatalytic systems.

Historically, the efforts in theory and simulation for catalysis in China have been much less significant than those in Europe and the United States. This may have been the result of either the significant expense for high-performance computing or the lack of specific training in this area. The availability and affordability of high-speed computing and electronic structure methods, however, have recently fueled a significant increase in the number of researchers and published papers in this area. In addition to the increased computational availability, there are a number of the younger scientists who have worked with some of the leading theory and catalysis groups in Europe and in the United States as postdoctoral fellows and have since returned to China as assistant professors. The efforts in theory and simulation in China have almost all been focused on the application of electronic structure methods to model metals, metal oxides, metal sulfides, and mesoporous catalytic materials, and the reactions they catalyze. Most of these efforts are focused on energy conversion processes. The scientists in computational catalysis are nearly all assistant professors. These scientists tend to be single investigators who are spread across China and tend to be less connected with experimental efforts. This is unlike the efforts in Europe and Japan in which theory and simulation are found in larger groups and are more directly tied to experiment. The WTEC panel found few efforts in Korea with the exception of the group of Professor Jae Sung Lee at Pohang University who was using theoretical calculations to complement experimental efforts in determining the optical properties of different photocatalytic materials.

5.4.3 *Comparison of Europe, Asia, and the United States*

If one compares the nature of the research, the experience of the researchers, and the infrastructure for research, there are some clear differences among Europe, Asia, and the United States. As discussed above, the research effort in Europe is well-established and has been carried out by distinguished scientists in Chemistry and Physics who have pioneered the development and application of *ab initio* methods to catalysis over the past three decades. This includes very strong efforts in Denmark, France, Germany, Italy, the Netherlands, Spain, and the United Kingdom. There has been an almost linear increase over time in the addition of young faculty in computational catalysis across Europe in order to offset current and near-future retirements. There are a few exceptions, however, such as at the Technical University of Denmark, where a significant grant from a private foundation has helped to establish a very strong program in the computational design of catalytic materials and to hire outstanding young faculty.

The efforts in Japan are focused in just a few institutions and are run by well-established faculty. While there are strong efforts in method development, such as the development of novel QM/MD simulations and combinatorial methods, most of the work is focused on the application of theory and simulation to different catalytic systems with strong programs in photocatalysis.

In contrast, computational catalysis in China is increasing exponentially and is comprised universally of young faculty who are dispersed throughout China. The computational programs, however, do not appear to be very well connected with the experimental programs.

The efforts in theory and simulation in the United States have grown significantly over the past 15 years. While there have been a number of earlier pioneers in the field in chemistry, the efforts were perhaps more fragmented as programs stretched across a range of material systems. In addition, relatively few of the many young theoretical or computational chemists hired into chemistry or physics departments have focused on heterogeneous catalysis. Their efforts have focused instead on novel theoretical and computational method developments, drug design, enzyme chemistry, and other biological systems. This may be due to the fact that catalysis is typically viewed in the United States as a technology rather than as a basic science. As a result, there has been much more funding for biomaterials and biocatalysis than for heterogeneous catalysis. Although the number of researchers focused on heterogeneous catalysis is smaller than the number focused on biocatalysis, those in the field have been instrumental in developing novel electronic structure algorithms and modeling for both homogeneous as well as heterogeneous catalysis.

While over the past decade there have been few new faculty members in Physics and Chemistry hired in the United States to work in computational heterogeneous catalysis, there has been an exponential growth in young faculty hired into Chemical Engineering over this time period. This is perhaps a natural result of a continuing shift in Chemical Engineering to move from understanding classic macroscopic behavior to the elucidation of fundamental molecular-level science. In terms of modeling, there has been a shift from continuum-level modeling to molecular-level simulations. Molecular simulation has been a core strength in US Chemical Engineering. This shift to more detailed simulations to establish the fundamental science has now begun to move into simulating electronic structure and its influence on catalytic behavior. The emphasis has been focused on the application of *ab initio* methods.

The development of electronic structure methods, however, requires a large concerted effort, with a strong basis in solid-state physics. This creates a considerable challenge for young assistant Chemical Engineering professors, who are required to show significant productivity in their first few years in order to receive tenure. The electronic structure codes that are used today have been developed over a number of years by a strong concerted commitment. The funding and tenure situation in the United States makes the development of such electronic structure codes in Chemical Engineering departments much more difficult. The United States, however, has traditionally pioneered many of the leading efforts in atomistic and molecular-level simulations in catalysis. The strong foundations of molecular thermodynamics and kinetics that are taught in the US Chemical Engineering curriculum is perhaps the major driving force for this. There has been a shift, however, over the past decade from atomistic-scale simulation development efforts to the application of *ab initio* methods. The application of both electronic structure methods and simulations has changed over the past decade as the focus of the funding agencies has changed. Most of the current efforts have targeted energy. This includes efforts in gas to liquids, electrocatalysis for fuel cells, conversion of biorenewables, and photocatalysis.

5.5 Universal Trends

5.5.1 *The good news*

The assessment from the WTEC site visits together with an analysis of the recent literature revealed some clear universal trends across the field of computational catalysis:

- Strong coupling between theory and experiment
- Simulation of more realistic reaction environments
- Isolation of factors difficult or impossible to achieve experimentally
- Applications moving to energy-related issues
- Slow improvement in method accuracy
- Invaluable partnership of theory and experiment.

These trends highlight some of the important advances that theory and simulation have made over the past decade. For example, it is now possible to provide structural information such as bond lengths and angles concerning the surface structure or molecules bound the surface to within 0.05 Å and 2°, respectively, and spectroscopic signatures of adsorbed intermediates to within a few percent of experiment (Forseman and Frisch, 1996; van Santen and Neurock, 2006). Similarly, theory can readily determine adsorption energies, reaction energies, as well as activation barriers for important elementary steps within reasonable time frames (1–2 days) and accuracies (< 20 kJ/mol), which provides a strong framework for elucidating reaction mechanisms. Theory has thus become an invaluable partner with experiment in understanding catalytic reaction systems. In addition, there appears to be a growing trend around the world where theory and simulation are being used to help guide the search for new materials.

The advances in coupling theory and experiment have come from improved computational abilities as well as a greater acceptance of theory and simulation in the experimental and the industrial communities. The exponential advances in raw computational speed, increased memory, improved algorithm development, and massively parallel architectures have allowed simulations to move to much larger systems and more realistic reaction environments in reasonable time frames. Much of the theoretical work up until this past decade was focused on the chemisorption and reactivity of small molecules (< 5 atoms) on ideal single crystal surfaces carried out at very low coverages. This work provided a wealth of useful information and aided ultrahigh-vacuum surface science studies but offered few direct insights into catalysis at operating conditions. Advances that have occurred over the past few years, however, have made it possible to follow the reaction chemistry as well as a reliable portion of the local catalytic reaction environment, which includes more realistic metal clusters, metal-support interactions, higher surface coverages, the influence of alloying, solution effects, and even the presence of electric fields.

Theory in particular is being used to elucidate properties that would be difficult, and in a number of cases impossible, to determine experimentally in order to provide insights as to what controls intrinsic reactivity and catalytic chemistry. There are increasing efforts aimed at interrogating the electronic, entropic, and intramolecular interactions that control the nature of the transition states and reactive intermediates for specific molecules.

In addition to the advances in system size and the time required to carry out particular simulations, there have also been important improvements in the methods, which have helped to improve method accuracy. For example, there are new exchange-correlation functionals that provide a more exact accounting of the exchange integrals as well as the implementation of many-body perturbation theory, namely GW theory (Hedin, 1965; Hedin and Lundqvist, 1969; Hybertsen and Louie, 1986; Hafner, 2008) to calculate self interactions and the screened Coulomb potential. This provided more accurate predictions of band gaps and excited state properties for particular materials, as well as more accurate ground state properties. In addition, higher level *ab initio* methods are being implemented that can explore much more complex surfaces as well as the nature of the reaction centers more rapidly and thus allow for significantly higher accuracies.

It was clear from the WTEC site visits as well as what is present in the current literature that theory and experiment are increasingly valuable partners and offer powerful contributions to help determine reaction intermediates, elucidate the nature of active sites, establish reaction mechanisms, and search for new materials.

5.5.2 *The not-so-good news*

Despite the significant advances and the exponential increase in the use of theory and simulation, there are still a number of important limitations. Some of those established in the WTEC assessment are given here:

- Method accuracy is still an issue
- Connection to actual kinetics is still weak
- Few efforts exist to link *ab initio* methods and atomistic simulations necessary to describe coupled adsorption, diffusion, and reaction
- Simulating photoexcitation and photocatalysis is a difficult challenge
- A significant shift has occurred in emphasis, from development to application.

There have been a number of advances in theory, which include the development of new exchange-correlation functionals for density functional theory (Heyd and Scuseria, 2003; Tao *et al.*, 2003; Perdew *et al.*, 2004; Staroverov *et al.*, 2004; Perdew *et al.*, 2005; Csonka *et al.*, 2007; Janesko and Scuseria, 2007; Tao *et al.*, 2007; Hafner, 2008; Tao *et al.*, 2008), novel embedding schemes (Govind *et al.*, 1999; Sauer and Sierka, 2000; Vreven and Morokuma, 2000), and many-body perturbation methods for surfaces (Hafner, 2008). Despite these advances, the accuracy in predicting adsorption energies, reaction energies, and activation barriers is still 0.2 eV with known pathological outliers. This prevents accurate kinetic predictions, which require 0.05 eV or less. There are also no systematic ways known to improve the accuracy of density functional theory, which provides the foundation of most of the work in theory for catalysis.

While there has also been a significant increase in the application of theory to catalysis, most of the work has been focused on the calculation of single elementary steps and their energies, with little regard for the myriad simultaneous kinetic processes or the operative catalytic kinetics. Speculations on the steps that control catalysis are often made from a simple idealized potential energy surface without concern for the surface coverages, reaction rates, and competing pathways.

In many catalytic systems, adsorption, diffusion, and reaction are intimately coupled and can control the catalytic performance. This is particularly true in zeolites. This requires both *ab initio* methods and atomistic simulations and the ability to seamlessly integrate these over different time and length scales. While there is some effort to determine the kinetics of adsorption, reaction, and diffusion, each of these tends to be addressed separately. As such, the solution strategies tend to be ad hoc. The appropriate treatment will likely require the development of new hybrid or coarse-grained methods that can appropriately treat different length and time scales.

As discussed in Chapter 8, Applications of Catalysis: Renewable Fuels and Chemicals, there is an exponential rise in the focus on photocatalysis with response to our future energy challenges. Despite the increase in experimental efforts, the theoretical efforts in photocatalysis have been much more limited. This is due to the challenges of simulating the chemistry of excited states. A rigorous approach to simulating photoexcited states and photocatalysis would require the ability to follow time-dependent processes, and thus the dynamics of intermediates that form, upon a changing excited-state potential energy surface and the cascade of different products. This cannot be solved using traditional static *ab initio* methods, because electron transfer processes and state-to-state transitions are dynamic. While there have been a number of tremendous advances in the application of time-dependent methods such as time-dependent density functional theory, they have been limited to following the reaction dynamics of very small systems (Serrano-Andres and Merchan, 2005).

Finally, it is important to note that there has been a significant shift in the emphasis of theory efforts from development to application. There has been an exponential rise in the number of researchers in China and the United States who have recently moved into theory and simulation for catalysis, most of whom have been focused on the use of theory and simulation to examine particular catalytic problems. This is likely linked to the background of the researchers as well as to the nature of funding in the United States and China compared to that in Europe and Japan. The significant increase in the number of scientists in the United States has been in the area of chemical engineering, where the focus has been on solving important catalytic problems. This is driven to some extent by the funding agencies and perhaps by societal views that catalysis (in general) is more of a technology than a fundamental science. Development of theoretical or simulation methods that will significantly enhance our understanding of catalysis are thus only funded when they are tied to the solution of current technological issues such as energy conversion and environmental remediation. While the application of theory is important, there is still a strong need for the method development that will more broadly aid our ability to simulate catalytic systems. This tends to be an issue in the United States as well as China, and it is of growing concern in Europe as well.

5.6 Examples of Applications of Theory and Simulation

Computational catalysis has become an important tool in interpreting both spectral and structural information, resolving the reaction intermediates, elucidating the nature of the active site, determining the reaction kinetics, and aiding the design of new materials. Below are highlighted some of the important contributions that theory and simulation have made to the following areas:

- Connecting theory and spectroscopy
- Modeling more realistic reaction environments
- Applications to energy
- Simulating catalytic performance
- Design in heterogeneous catalysis
- From theory to synthesis.

5.6.1 *Connecting theory and spectroscopy*

The strongest coupling between theory and experiment has been in the use of theory to predict, interpret, and resolve the structural as well as spectroscopic signatures concerning active surface sites and the nature of the reactive surface intermediates. Theory is being used routinely to simulate infrared, Raman, nuclear magnetic resonance, ultraviolet-visible, X-ray absorption, X-ray photoemission, near-edge absorption fine structure, as well as scanning tunneling and atomic force microscopy in an effort to complement, confirm, and help explain experimental results.

An elegant example of this was shown in the work by Sautet's group (Haubrich *et al.*, 2006; 2008), which combined extensive density functional theoretical calculations, high-resolution electron energy-loss spectroscopy (HREELS), temperature program desorption spectroscopy, and low-energy electron diffraction in order to identify all of the possible modes of adsorption and the thermal decomposition of prenal (3-methyl-2-butenal), a model α, β unsaturated aldehyde, on Pt and Pt-Sn alloy surfaces. DFT calculations were able to specifically identify the five unique modes of adsorption for prenal (η^2-diσ(CC)-s-trans, η^2-diσ(CC)-s-cis, η^3-diσ(CC)-σ(O)-cis, η^4-diσ(CC)-diσ(CO)-s-trans and η^4-π(CC)-diσ(CO-s-cis) and predict the vibrational frequencies for each of these modes. While the adsorption energies were found to be quite similar for specific surfaces, the simulated HREELS spectra provided a direct fingerprint that could uniquely identify all the modes in the complex experimental HREELS spectra. All five modes of adsorption appear to be present on the Pt(111) surface, demonstrating adsorption energies that range from −47 to −59 kJ/mol. All five adsorption modes along with their adsorption energies are shown in Figure 5.3. A comparison of the computed and experimental HREELS spectrum at

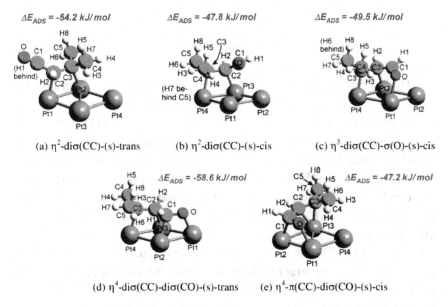

Fig. 5.3. DFT-predicted lowest energy adsorption states for prenal (3-methyl-2-butenal) bound to Pt(111) (Haubrich et al., 2008).

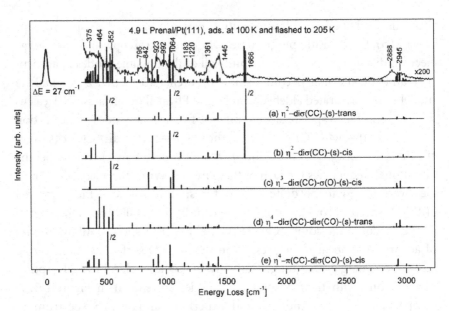

Fig. 5.4. Comparison of the experimental and theoretically predicted HREELS spectra for 4.9 Langmuirs of prenal adsorbed onto Pt(111) at 205 K. The HREELS spectra for the five lowest energy adsorption states are given in examples (a)–(e) (Haubrich et al., 2008).

4.9 Langmuir of prenal on Pt(111), given in Figure 5.4, shows the impressive match between theory and experiment.

The adsorption of prenal on the Pt-Sn surfaces was found to be significantly different than that on Pt(111), as the adsorption energies were reduced to −39.1 and −30.8 kJ/mol on the $Pt_3Sn/Pt(111)$ surface and −33.4 kJ/mol on the Pt_2Sn surface. The most stable surface adsorption modes for prenal on the $Pt_3Sn/Pt(111)$ surface are shown in Figure 5.5.

A comparison of the experimental and theoretical results for the three most stable low-temperature adsorption modes on the $(\sqrt{3} \times \sqrt{3})R30°$ $Pt_3Sn/Pt(111)$ surface alloy is shown in Figure 5.6. The results reveal that after annealing the surface to 200 K, only the η^1 mode appears in the HREELS spectra. The predominant adsorption structures that lie parallel to the surface require larger surface ensembles that are not present on the alloy surface. This is noted by the absence of the characteristic vibration frequencies at 375, 956, 1371, and 1444 cm^{-1} in the HREELS spectra. The addition of Sn shuts down the larger Pt ensembles, thus preventing the

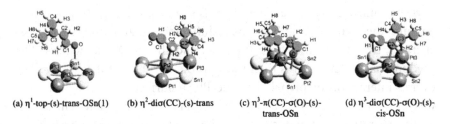

(a) η¹-top-(s)-trans-OSn(1)   (b) η²-diσ(CC)-(s)-trans   (c) η³-π(CC)-σ(O)-(s)-trans-OSn   (d) η³-diσ(CC)-σ(O)-(s)-cis-OSn

Fig. 5.5. DFT predicted lowest energy adsorption states for prenal (3-methyl-2-butenal) bound to Pt$_3$Sn/Pt(111) (Haubrich et al., 2008).

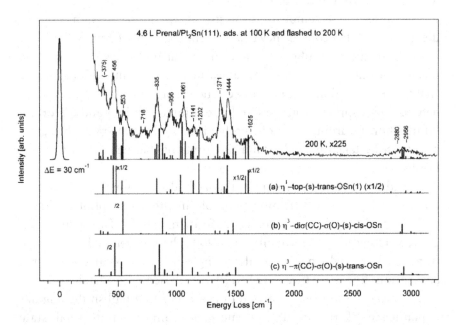

Fig. 5.6. Comparison of the experimental and the theoretically predicted HREELS spectra for 4.6 Langmuirs of prenal adsorbed onto Pt$_3$Sn/Pt(111) at 205 K. The HREELS spectra for the 3 lowest energy adsorption states are given in (a)–(c) (Haubrich et al., 2008).

higher-fold η^2 and η^4 adsorption modes. The interaction between prenal and the surface becomes significantly weaker, as the prenal now adsorbs through its oxygen in an η^1 mode atop of the Sn sites. The results from theory and experiment both show that the addition of Sn leads to a more weakly held prenal intermediate, which binds atop of Sn through its oxygen atom. Theory reveals that there is significant charge transfer from Sn into the neighboring Pt, which shifts the d-band on Pt to lower energies, thus weakening both the Pt-C and the Pt-O interactions. The predominant interaction is then through the Sn-O bond that forms. The overall interaction of prenal with the Pt-Sn surface is thus significantly weaker than the prenal interaction with Pt(111). As such, there is little shift of the vibrational frequencies to those found in the gas phase. DFT, however, cannot readily discriminate between these weak adsorption modes, because they are governed by weak dispersion forces. This begins to push the current limits of theory.

The results here demonstrate the strong synergy achieved in coupling theory and spectroscopy as they enable complex surface compositions that result from the adsorption of multifunctional molecules on both pure metals and alloys to be resolved at various temperatures. This should enable the resolution of the complex molecular interactions that occur between co-adsorbed molecules under reaction conditions and important information regarding reaction mechanism. It also illustrates that great care must be taken to understand the limits of the theory and where it breaks down.

In addition to the coupling of theory and vibrational spectroscopy, theory is also playing an important role in predicting photoemission and X-ray absorption spectroscopy. Hermann *et al.*, (Hermann *et al.*, 2001; Kolczewski and Hermann, 2003; 2004) have developed an *ab initio* framework by which to simulate angle-resolved, near-edge X-ray absorption fine structure (NEXAFS) spectroscopy. A detailed set of *ab initio* calculations were carried out in order to establish the angular dependence of the excitation energies from different final state molecular orbitals along with the dipole transition matrix elements. This information was subsequently used to construct angle-resolved NEXAFS spectra for different electronic states of individual atoms in the surface and help resolve the O 1s excitation for the three different types of oxygen atoms in the V_2O_5(010) surface. The three different coordination environments for oxygen (1-fold, 2-fold, and 3-fold) result in different excitation and absorption spectra that overlap, thus making it difficult, and in some cases impossible, to resolve from experiment alone.

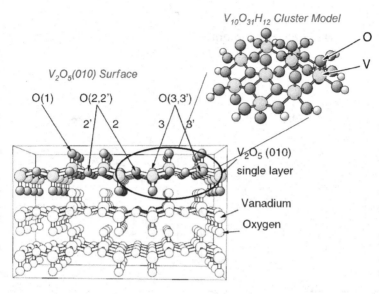

Fig. 5.7. Model of the V_2O_5 (010) surface and the $V_{10}O_{31}H_{12}$ cluster model used to calculate the NEXAFS spectra (Kolczewski and Hermann, 2003).

A comparison between the *ab initio* simulated NEXAFS spectra from the $V_{10}O_{31}H_{12}$ model cluster cut from the $V_2O_5(010)$ surface (Figure 5.7) and the experimentally measured $V_2O_5(010)$ surface spectra (Figure 5.8) is shown for both xz polarization and with two different polar angles $\varphi_{pol} = 90°$ and $20°$. The agreement between theory and experiment for the overall spectra is remarkably good. Theory is able to ultimately deconvolute the two major peaks in the spectra into the specific individual excitations from the oxygen states that are responsible. Results reveal that the two peaks at 530 and 535 eV are the result of excitation from the O 1s core electrons to final state orbitals that are comprised of antibonding O 2p and V 3d states. Figure 5.9 shows the orbitals.

The states that appear in the NEXAFS spectra are the result of different local bonding environments and the different oxygen coordination. As such, one can directly assign the two excitation peaks to specific oxygen atoms. The results in Figure 5.8 reveal that for x polarization, both oxygen peaks reveal strong contributions from singly coordinated oxygen atoms,

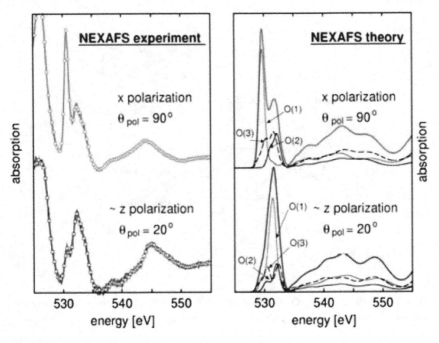

Fig. 5.8. Comparison of (*left*) experimental NEXAFS spectra and (*right*) theoretically-predicted O 1s core level excitation spectra for different polarization angles (Kolczewski and Hermann, 2003).

whereas for y polarization the high-energy peak contains contributions from the 3-fold oxygen atoms.

This combination of theory with experiment should prove to be extremely valuable, because it provides for the direct resolution of the states and the specific structures that give rise to specific X-ray absorption peaks. The interrogation of the surface states before and after adsorption and reaction should enable one to distinguish which of the states are involved in the chemistry and the specific nature of the coordination environment.

The two systems discussed above are just two recent examples of the universal trend in which theory and spectroscopy are being combined to provide unprecedented resolution of the structural features of the catalyst and the surface intermediates under reaction conditions.

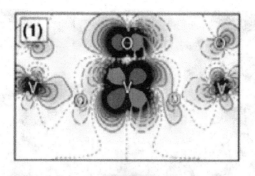
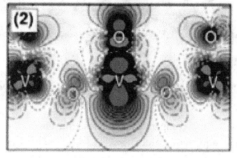

Fig. 5.9. Selected final state orbitals for O(1) 1s core excitations (Kolczewski and Hermann, 2003).

5.6.2 Modeling more realistic reaction environments

A second major trend that has taken place over the past few years is the move from the ideal low-coverage single-crystal surfaces to more realistic reaction environments. Simulating the specific conditions that occur *in situ* at the surface of an actual catalyst presents a number of significant challenges, due to the complexity of the environment around the active site. Fully elucidating the reactivity of a supported metal particle, for example, would require an understanding of the metal-support interactions; particle size effects; the influence of alloying; the role of defect sites, promoters, or poisons; and the influence of solution if present. This is shown schematically in Figure 5.10. While we cannot currently model such a complex situation, there has been an increasing number of theoretical and simulation studies aimed at elucidating some of these effects and their importance in catalysis. These results will not only aid in

Fig. 5.10. Schematic of the complex catalytic reaction environment for a supported bimetallic particle in an aqueous medium.

the development of structure-property relationships but also help to establish more coarse-grained models which can mimic these effects. We present two examples which help to highlight the effect of the reaction environment. A more detailed discussion is also given in an example presented later on simulating catalytic performance.

The increases in computing power that have taken place since 2003 have allowed for the simulation of much larger system sizes to be explored over reasonable time frames. This permits the examination of more extended and realistic features of the reaction environment. In particular, there have been a growing number of studies that have focused on the influence of the support on the activity of metal. This has been fueled in particular by the fascination of the community as to why supported Au nanoparticles are active for a wide range of reactions when bulk gold itself is inactive. The support is thought to play an important role. In order to understand metal-support interactions and their role in catalysis, ultrahigh-vacuum studies have been carried out over nanometer metal particles and layers deposited onto ideal model metal oxide surfaces grown on top of a metal substrate. Previous studies carried out over these supported Au clusters suggest that anionic Au formed as a result of charge transfer from the oxide to the metal is responsible for the unique catalytic activity for the oxidation of CO. Recent combined theoretical and experimental studies by Pacchioni and Freund (Sterrer *et al.*, 2006; Yulikov *et al.*, 2006; Sterrer *et al.*, 2007; Giordano *et al.*, 2008; Ulrich *et al.*, 2008) have shown that the thickness of the oxide films supported on metal substrates plays an important role in the charge transfer properties of the oxide interface that sits between the metal and the support.

Theoretical results from Pacchioni and colleagues (Sterrer *et al.*, 2006; Yulikov *et al.*, 2006) suggest that the charging that occurs may not be associated with the Au/MgO interaction but instead be the result of a charge transfer from the metal substrate through the thin MgO layer and into the Au catalyst particles. They suggested that the thickness of the oxide layer might be ultimately tuned in order to control reactivity. To confirm this, they carried out careful scanning tunneling microscopy (STM) studies of Au and Pd deposited onto 3 ML MgO films supported on Au(001). The STM studies of the surface reveal that Au deposits as single adatoms and order on the thin MgO support with well-defined nearest-neighbor distances, whereas Pd forms a more random dispersion (Yulikov *et al.*, 2006), as shown in Figure 5.11. This indicates that the electronic structure plays an important role in controlling nucleation.

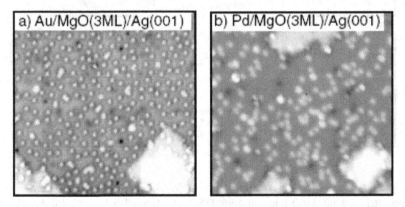

Fig. 5.11. STM images for Au (a) and Pd (b) supported on 3 mL of MgO grown on Ag(001) (Sterrer *et al.*, 2007).

Pacchioni and colleagues demonstrate that the ordering of Au on the oxide is the result of repulsive interactions between the Au adatoms on the surface; these repulsive interactions are not present, however, for Pd. Detailed density functional theory calculations using the Tersoff–Hamann approximation for the electronic structure were used to simulate the STM images of neutral and charged Au and Pd atoms on an MgO film (Yulikov *et al.*, 2006). The results were compared with experimental STM images to help resolve the charge states of the metal atoms. The simulation results for anion and the neutral Pd atoms provide a direct match with experiment, as is shown in Figure 5.12. For Au, both the simulation and the experimental STM results reveal a protrusion that is followed by a depression for anionic supported Au adatoms. The results for supported Pd atoms simply show a diffuse protrusion. The "Sombrero" protrusion-depression effect only occurs, however, when Au is charged negatively.

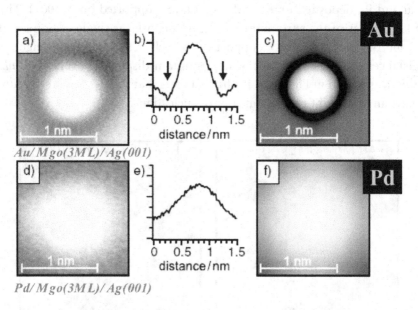

Fig. 5.12. Comparison of experimental STM images for Au (a) and Pd (d) atoms supported on 3 mL MgO films on Ag(001) with DFT-simulated images for Au (c) and Pd (f) on 3 mL MgO supported on Ag(001). The experimental height profiles on Au and Pd are given in (b) and (e), respectively (Sterrer *et al.*, 2007).

Metallic Au displays only a protrusion. In other studies Pacchioni et al., have demonstrated that Au can become anionically charged when it is bound to color centers on the oxide (Sterrer et al., 2006); Au bound to terraces remained neutral. The adsorption of CO onto the anionic supported Au provides a clear fingerprint in the CO vibrational stretching frequencies by which one can analyze the influence of the support.

The results here show the important synergy between theory and experiment and help to demonstrate that the charge transfer that exists between the metal and the support is a function of the oxide layer thickness as well as a function of the nature of the supported metal.

In addition to the support, the effect of the gas phase conditions at the active site can also be important in controlling the site's activity and selectivity. The gas phase pressures and the surface kinetics ultimately control the surface coverages, which in turn dictate the rates of reaction. Scheffler and colleagues developed a novel first-principles approach to determining the thermodynamically preferred surfaces at different conditions (Scheffler, 1988; Reuter and Scheffler, 2002; Stampfl et al., 2002; Reuter and Scheffler, 2003b). This approach has now become the accepted standard in establishing the thermodynamically favored surface compositions for metals, metal oxide, and metal sulfide surfaces. The calculation of free energies for different surface structures are used together with the chemical potentials for the gas phase intermediates to calculate the lowest free energy surface states that can form at specific partial pressures and temperatures.

Raybaud et al. (Arrouvel et al., 2005; Costa et al., 2007; Raybaud, 2007; Raybaud et al., 2008), extended the idea to understanding the surface composition and the potential reactivity of metal sulfide particles and their interaction with different supports. More specifically, they examined the metal-sulfide particle formation on both an anatase TiO_2 and a γ-alumina support (Costa et al., 2007). The partial pressures of water, hydrogen, and H_2S at different temperatures were used together with DFT-calculated surface free energies to determine the hydroxylation and sulfidation state of the support under different reaction conditions (Arrouvel et al., 2005; Costa et al., 2007; Raybaud, 2007; Raybaud et al., 2008). DFT calculations on the Mo and sulfur edges of the different MoS_2 and CoMoS surfaces were used to understand the reactivity of these supported surfaces and

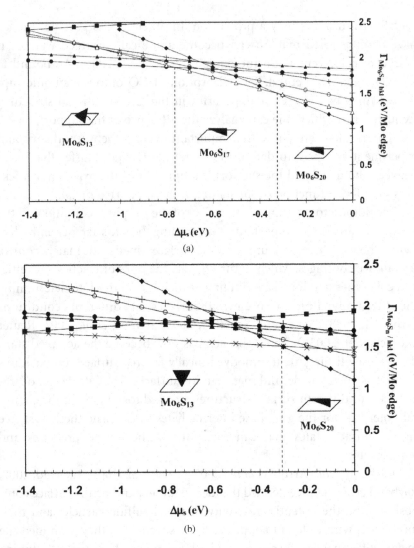

Fig. 5.13. DFT-calculated surface phase diagrams for Mo_6S_{13}, Mo_6S_{17}, Mo_6S_{20} clusters supported on (a) gamma-Al_2O_3 (100) and (b) anatase TiO_2(101) (Costa et al., 2007).

particles. The surface phase diagrams for Mo_6S_x clusters on γ-alumina and anatase TiO_2(101) surfaces are shown in Figure 5.13.

The results reveal characteristically different wetting behavior for MoS_2 clusters supported on anatase TiO_2 than those on γ-alumina. The

MoS$_2$ clusters prefer epitaxial growth on anatase TiO$_2$, which leads to the formation of tilted structures with respect to the support, and therefore, to poor wetting of the support. These same clusters on the γ-alumina resulted in much better wetting of the support. The surface free energies of the S-edge and the Mo-edge of the particles anchored to TiO$_2$ and γ-alumina were subsequently used together with a Gibbs–Curie–Wulff–Kaischew method to determine the geometric parameters of the supported MoS$_2$ clusters (Costa et al., 2007). The resulting proportion of free Mo and S edge sites as a function of the particle diameter on both anatase TiO$_2$ and γ-alumina is shown in Figure 5.14. The average particle size for MoS$_2$ on TiO$_2$ is about 38 Å, whereas that on γ-alumina is higher at 49 Å.

The results in Figure 5.14 would then suggest that there are a similar number of free edge sites. The relative percentage of S/Mo edge sites however was found to be higher for γ-alumina. There is a loss of S-edges on anatase due to edge wetting of the support. In addition, there appears to be a stabilization of particles with higher vacancies on the TiO$_2$ support. As a result, the higher intrinsic activity for HDS on anatase over γ-alumina is due to these differences in the nature of the exposed edge sites.

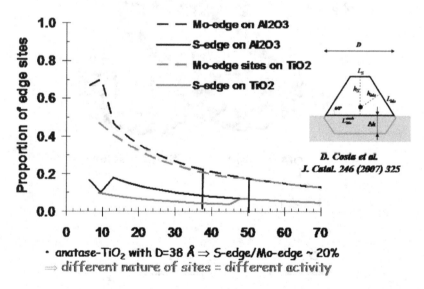

Fig. 5.14. Distribution of sulfur and molybdenum edge sites as a function of MoS$_2$ particle diameter supported on γ-alumina and TiO$_2$ anatase substrates (Costa et al., 2007).

Recent advances in embedding methods have allowed for the resolution of structure and reactivity in complex micro- and mesoporous materials. The region about the active site is treated quantum-mechanically, whereas the effects of the porous framework are treated with lower-level QM or molecular mechanics (MM) models. Malek, Li, and van Santen (Malek et al., 2007; Zhang et al., 2008), for example, used a DFT/MM embedding approach to follow the potential energy surface for the enantioselective epoxidation of cis- and trans- methyl styrene by an oxo-Mn-V-salen complex anchored into the channels of MCM 41.

The optimized structure of the anchored complex is shown in Figure 5.15. Both electronic and steric confinement effects were found to cooperatively control the reactivity and enantioselective behavior of this material. The triplet state potential energy surface for the oxygen transfer reaction to the adsorbed trans-methyl styrene molecule is shown in

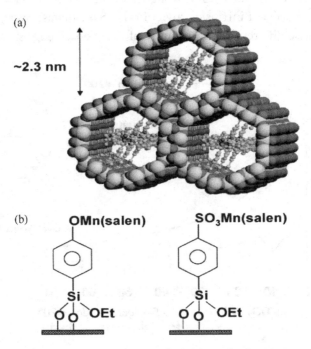

Fig. 5.15. Mn-salen complexes anchored inside the channels of MCM-41 and the explicit anchors used in the DFT calculations (Malek et al., 2007).

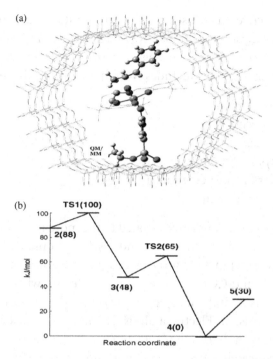

Fig. 5.16. (a) The DFT/MM optimized adsorbed methylstyrene onto the Mn-salen complex anchored inside the MCM-41 channel, (b) the reaction energy diagram for the oxygen transfer reaction along the triplet energy surface (Malek et al., 2007).

Figure 5.16. Even though trans-methylstyrene was found to have the greatest asymmetric induction effect, the cis form resulted in the lowest activation barriers. The immobilization by the electron-donating linker used to anchor the complex was found to decrease the barriers. In addition, the channel structure was also found to reduce the activation barriers and enhance enantioselectivity as the result of the confinement and distortion of the Mn-salen complex. Channel confinement was therefore found to greatly depend upon the channel size.

5.6.3 *Applications to energy*

The application of theory and simulation to catalysis has covered a broad range of different applications, including NO_x reduction, selective oxidation, selective/enantioselective hydrogenation, dehydrogenation, catalytic cracking, isomerization, reforming, desulfurization, and dechlorination. The recent trends, however, reveal an increasing focus on the application to energy. This includes work in the following areas:

- Methane conversion
- Fischer–Tropsch synthesis
- Electrocatalysis
- Photocatalysis (just emerging)
- Coal conversion (just emerging)
- Biorenewables (just emerging).

Most of the energy efforts are related to the more traditional processes such as methane conversion routes and Fischer–Tropsch (FT) synthesis. This is due to the fact that the initial reactants for methane conversion and FT, namely CH_4 and CO/H_2, are small molecules for which there is considerable experimental data and previous theoretical results on well-defined surfaces. Electrocatalysis is an area that has witnessed tremendous growth since about 2002. The challenges of modeling the complex electrocatalytic reaction environment present in a proton-exchange membrane fuel cell, however, have severely limited most modeling efforts. A rigorous description of the reaction environment requires simulating the metal, carbon support, polymer electrolyte, solution phase, and constant electrochemical potential. Despite this complexity, there have been a number of initial successes in modeling electrocatalysis using simple models of the surface and the solution. These initial successes as well as the importance of fuel cells have stimulated a strong and recent interest in modeling these systems. The small size of the molecules of interest (CO, H_2, O_2, H_2O, CH_3OH, HCOOH, etc.) also helps to make electrocatalysis more attractive for theoretical studies.

There is also a growing interest in modeling other more complicated systems such as photocatalysis, coal conversion, and biorenewables. There is a concerted effort in Japan, for example, to model photocatalysis, and China has strong efforts in coal conversion. Despite the interest, the complexity of the reaction environment in all three of these areas has significantly inhibited progress; all three areas, however, will likely see considerable future growth for modeling in Asia, Europe, and the United States.

5.6.4 Simulating catalytic performance

While theory can be used to determine the elementary reaction steps involved in a catalytic system, the activation barriers for each step and the influence of the reaction environment, it cannot provide catalytic performance or kinetics directly. Catalytic activity and selectivity require simulating the rates of reaction, which are inherently dependent upon the rate constants as well as the surface coverages under working conditions. This can be carried out either via the development of microkinetic models or through the use of kinetic Monte Carlo simulations, as discussed earlier. A nice example of the latter was presented by Reuter and Scheffler, who combined *ab initio* calculations with rigorous statistical mechanics and integrated the results into kinetic Monte Carlo simulations to follow the oxidation of CO over the $RuO_2(110)$ surface (Reuter and Scheffler, 2002; 2003a; 2003b; 2006; Reuter et al., 2004).

A rigorous set of DFT calculations was used together with the chemical potentials for oxygen and Ru to determine the free energies for different terminations of the $RuO_2(110)$ surface as a function of the temperature and the partial pressures of oxygen and CO. The resulting surfaces that can form and their free energies are plotted with respect to the partial pressure of oxygen in Figure 5.17. These surfaces are comprised of Ru and oxygen centers. Oxygen and CO can adsorb at either the bridge (br) or the coordinatively unsaturated Ru sites (cus), referred to as O_{br} and O_{cus} and CO_{br} and CO_{cus}, respectively.

The density functional theory results reveal that as the partial pressure of oxygen is increased, the most stable surfaces shift from CO_{br}/CO_{cus} coverages at low oxygen partial pressures to O_{br}/CO_{cus} coverages at moderate pressures, and then to O_{br}/O_{cus} coverages at very high oxygen partial pressures, which might be expected as oxygen begins to compete for both the O_{br} and O_{cus} sites. The results were subsequently used to map out a thermodynamic phase diagram for the different states of oxygen and CO as a function of the partial pressures of CO and O_2, as well as the reaction temperature. The activation barriers for the CO oxidation at different sites and O_2 desorption were also calculated and used together with the transition state theory to determine the rate constants. The detailed reaction PES for the reaction of $CO_{cus} + O_{cus}$ is shown in Figure 5.18.

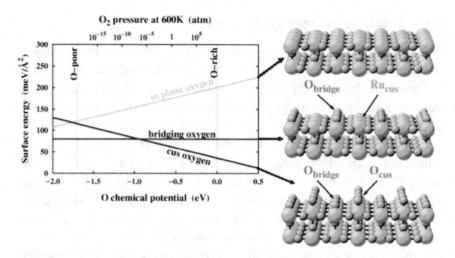

Fig. 5.17. DFT-derived phase diagram for oxygen on $RuO_2\{110\}$. The surface free energies for the in-plane oxygen, O_{br}/Ru_{cus} and O_{br}/O_{cus} terminated surfaces as a function of the chemical potential (and partial pressure) of oxygen at 600 K (Reuter and Scheffler, 2003b).

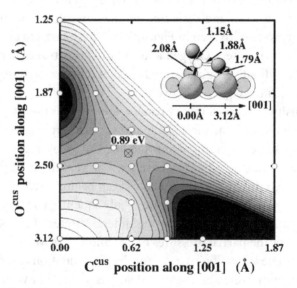

Fig. 5.18. DFT-calculated potential energy surface for the reaction of $CO_{cus} + O_{cus}$ on $RuO_2\{110\}$ to form CO_2 (Reuter and Scheffler, 2003a and 2003b).

The resulting *ab initio*-derived thermodynamic results along with the activation barriers were subsequently used as input to a kinetic Monte Carlo simulation to follow the temporal behavior for CO oxidation over the surface for a range of different reaction conditions. This allowed for the simulation of reaction rates from ultrahigh vacuum (UHV) conditions on up to high pressures of CO and O_2 and temperatures relevant to the actual catalytic conditions. The simulations provide a detailed understanding of the surface coverages, the rates, and the elementary steps that control the reaction at different conditions. At low partial pressures of CO, the simulations reveal that all of the coordinatively unsaturated Ru surface sites are covered with oxygen; CO is therefore blocked from the surface, which severely inhibits the rate. As the pressure of CO is increased up to about 20 atmospheres, the rate is considerably enhanced. Snapshots from the simulation reveal that CO is bound to the cus sites, whereas oxygen is predominantly located on the bridge sites. There is some mixing, though, as the fractional occupation of different sites at a particular point in time was found to be $N_{CO(br)} = 0.11$, $N_{CO(cus)} = 0.7$, $N_{O(br)} = 0.89$, and $N_{O(cus)} = 0.29$. The predominant reaction path to CO_2 involves the reaction $CO_{cus} + O_{cus} \rightarrow CO_2$. The reactions $CO_{br} + O_{cus} \rightarrow CO_2$ and $CO_{cus} + O_{br} \rightarrow CO_2$ still occur but are less favorable. An analysis of these paths demonstrates that an optimal balance is required between the lower activation barriers for the $CO_{br} + O_{cus}$ and the higher coverages of CO_{cus} and O_{br}. The results reveal the important message that a strict analysis of the activation barriers or the thermodynamics alone leads to incorrect conclusions on the steps that govern the kinetics. The accurate determination of the reaction rate thus requires the full kinetic simulation of all of the reaction steps, because they ultimately control the surface coverage as well as the reaction rate.

The highest CO_2 formation rates were found for systems with optimal CO and O_2 partial pressures that would result in surface compositions that lie within the very narrow region of the surface phase diagram where the CO_{br}/CO_{cus} and O_{br}/O_{cus} interfaces meet, as is seen in Figure 5.19. The dark regions in the right side of Figure 5.19 are those that have the highest activity. The simulation results were found to be in very good agreement with previously reported experimental results for CO oxidation over a range of different partial pressures of CO and O_2 as is shown in Figure 5.20 (Reuter and Scheffler, 2006). There has been some debate,

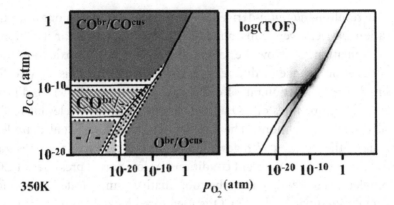

Fig. 5.19. First-principles DFT-based kinetic Monte Carlo simulation results for (*left*) the steady state surface structures of the RuO$_2$(110) surface, and (*right*) CO oxidation turnover frequencies over the RuO$_2$(110) surface as a function of the partial pressures of CO and oxygen at a temperature of 350 K (Reuter and Scheffler, 2006).

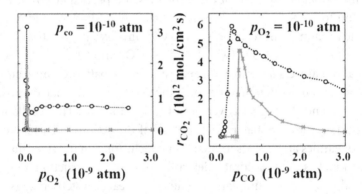

Fig. 5.20. Comparison of the first principle DFT based kinetic Monte Carlo simulation with experiments for the rate of CO$_2$ formation over RuO$_2$ at 350 K (Reuter et al., 2004).

however, as to whether the active surface is an oxide or a highly covered metal (Goodman et al., 2007).

The message here is not so much whether one can discriminate between the metal and oxide surfaces but instead the fact that the true understanding of catalysis requires a full consideration of the kinetics under working conditions and the influence of the appropriate steady state surface coverages. This was just one example of moving from a pure quantum mechanical treatment to simulating the actual kinetics under reaction conditions. While the simulation of kinetics is of critical importance to catalysis, there are still relatively few practitioners who attempt to connect *ab initio* results to reaction rates at appropriate conditions. Understanding the kinetics under working conditions will become even more important as the community shifts to more complex reaction systems.

5.6.5 *Design in heterogeneous catalysis*

In addition to the prediction of properties and the elucidation of mechanism, theory and simulation are beginning to be used to aid in the design of novel nanoscale catalytic materials. One of the key findings from the 2002 WTEC study on Applying Molecular and Materials Modeling (Westmoreland *et al.*, 2002) was the degree to which industry had used modeling to aid in homogeneous catalyst discovery and to protect intellectual property. There were very few efforts, if any, at the time aimed at guiding the design of heterogeneous catalytic systems. The complexity of heterogeneous catalytic systems was thought to be too great a challenge. Advances in modeling and detailed characterization over the past decade, however, have significantly increased our understanding of catalysis and have provided the basis for the emergence of various studies aimed at the design of new catalytic materials.

Most of the efforts to date have focused on the development of catalytic "descriptors" that can be used to predict experimental trends. This is analogous to the quantitative-structure-activity relationships used in drug design. One of the key differences, however, is that the electronic and structural descriptors used in drug design do not appear to work in predicting catalytic behavior. The development of "descriptors" that correlate the bulk behavior of metals, metal oxides, and metal sulfides with their reactivity has not been very successful. This may be due to the complexity of catalytic systems or due to the fact that bulk properties are not representative descriptors of the surface properties. More realistic descriptors might focus on properties that are tied to the actual bond-breaking and bond-making events that occur at the surface. Many of these steps appear to be governed by the strength of the adsorbate-surface bond. While this has been recognized for some time, it has only recently been used in order to screen catalytic materials as the increases in computational power now make it possible to readily examine a range of different materials.

This was elegantly demonstrated by the Nørskov group, which showed through detailed DFT calculations and a rigorous kinetic analysis that the optimal catalyst for the methanation reaction

$$CO + 3H_2 \rightarrow CH_4 + H_2O$$

would have an optimal CO dissociation energy (Andersson et al., 2006). This falls out of a Sabatier analysis that indicates the optimal material will facilitate the dissociation of CO yet still allow for both the carbon and the oxygen atoms that form to be hydrogenated on the surface. This is confirmed from DFT results for the elementary steps involved in the hydrogenation of CO to methane. As a result, the catalytic activity should follow the classic volcano shaped curve as a function of the calculated dissociation energies, as is shown in Figure 5.21. The lines in Figure 5.21 represent the best fit to the data. The optimal dissociation energy (ΔE_{diss}) appears to be 0.06 eV.

While Ni is used commercially, the results from the volcano curve in Figure 5.21 (bottom) reveal that Ru and Co are more active than Ni. Ru, however, is much more expensive than Ni. It is clear that the optimal material may need to satisfy more than a single design criterion, thus requiring the solution of a multi-objective optimization problem. Nørskov and colleagues used a Pareto-optimal set to guide their search for active low cost materials (Andersson et al., 2006). *Ab initio* DFT calculations were thus carried out together with a simple alloy interpolation model in order to screen 117 different alloys in the search for other likely alloy candidate materials. A Pareto plot of the results comparing the cost against the activity is shown in Figure 5.22. The best materials should be cheap and have a value for ΔE_{diss} as close to 0.06 eV as possible. The results show that while Fe is the cheapest material, its activity, as measured by the descriptor ΔE_{diss}, is not as good as Ni, Rh, Co, or Ru.

On the other hand, the costs of Ru and Rh are significantly higher than Ni, Co, and Fe. The optimal alloys are those that lie at "knee" in the bottom left of the Pareto plot, namely the $FeNi_3$ and NiFe alloys, because they

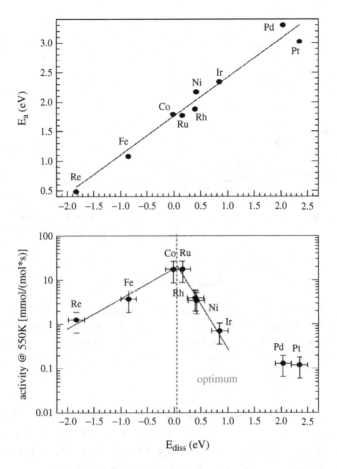

Fig. 5.21. The choice of an appropriate reactivity descriptor for the catalytic methanation reaction. (*Top*) The Brønsted–Evans–Polanyi relationship linearly relates the CO activation energy with the CO dissociation energy. (*Bottom*) CO activation over different metals shows a characteristic Sabatier principle with an optimum C–O bond dissociation energy (Andersson *et al.*, 2006).

have descriptors near optimal activity and are relatively low in cost. A class of different NiFe alloys was subsequently synthesized experimentally and tested (Kustov *et al.*, 2007). The results nicely confirmed the theoretical predictions, which indicate that the Ni_3Fe and NiFe alloys have activities that mimic those of Co but are significantly cheaper than Co.

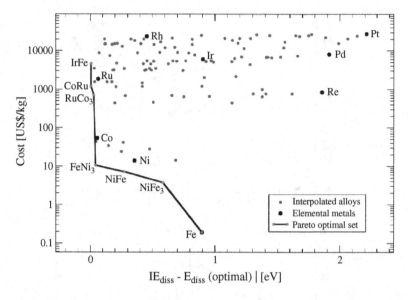

Fig. 5.22. The Pareto-optimal search for improved methanation catalysts. The optimal material should have the highest activity as measured here by the CO dissociation energy predictor along with the lowest cost. The best materials sit in the bottom left hand corner of the curve (Andersson et al., 2006).

The authors have used similar approaches to identify novel catalytic alloy materials for ammonia synthesis (Jacobsen et al., 2001; Logadottir and Nørskov, 2003; Bligaard et al., 2004; Boisen et al., 2005; Honkala et al., 2005; Hellman et al., 2006a; Hellman et al., 2006b) and, recently, acetylene hydrogenation (Studt et al., 2008). They conclude that the predictive power of DFT is attributed to the fact that the width of the volcano curve is large enough to overcome any systematic errors in calculating reaction energies. These are just few examples of the emerging trend of using theory to determine appropriate descriptors that can subsequently be used to screen different materials in the search of optimal catalysts.

5.6.6 *From theory to synthesis*

One of the key conclusions from a 2003 NSF workshop report *Future Directions in Catalysis: Structures that Function on the Nanoscale* (Davis and Tilley, 2003) was the need for theory and simulation methodologies to provide: "... deeper understanding and more complete descriptions of complex reactions and collective behavior including self-assembly in solution". It was recognized that simulating catalyst synthesis presents a tremendous but important challenge for theory and simulation, because these processes are controlled by ill-defined wet chemical laboratory methods. Despite its importance, little has been done in the area of synthesis. This is due to the complexity of modeling collective solution phase processes that govern synthesis, such as nucleation, precipitation, and growth. In the present assessment, all of the work that the WTEC panelists saw was focused on moving from structure to function rather than from synthesis to structure. There are, however, a few efforts in the literature that should be mentioned that have been aimed at the design of zeolite templates and their role in controlling the zeolite morphology.

Early efforts by Catlow and Thomas (Lewis *et al.*, 1995; 1996; Willock *et al.*, 1997; Catlow *et al.*, 1998) demonstrated how molecular-level simulations could be used to design template molecules to help direct the synthesis of specific zeolite frameworks. The templates are generally organic bases that are added during synthesis to control the crystallization of the porous framework structure around them. Simulations were thus used to "grow" the optimal template molecule inside a particular zeolite so as to optimize the interactions between the molecule and the pore. In this work, a new code termed ZEBEDDE (Zeolites by Evolutionary Denovo Design) was developed and used to carry out the following functions: (1) molecular building, (2) rotate, (3) shake and (4) rocking functions, as well as (5) bond twist, (6) ring formation, and (7) energy minimization in order to grow the molecule within the framework (Lewis *et al.*, 1996). This was demonstrated with the design of the 4-piperidiopiperidene template molecule to aid in the synthesis of the levyne zeolite structure. The intermediates generated in the "growth" process, starting with methane, are shown in Figure 5.23, along with the fit of the final 4 piperinopiperidene template within the levyne structure. While this work initially generated much excitement, there have been only a few subsequent papers in the area (Barrett *et al.*, 1996; Lewis *et al.*, 1997a; Lewis *et al.*, 1997b).

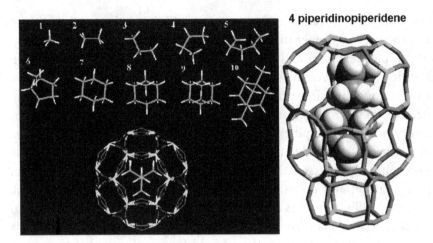

Fig. 5.23. The 4-piperidiopiperidene template molecule was established by using a *de novo* design strategy where atomistic simulations were used to "grow" the optimum molecule within the zeolite cage. The simulations start with methane shown in number 1 above (*top left*) and sequentially "grow" the template molecule within the microporous levyne structure via a sequence of different steps, including (1) build, (2) rotate, (3) shake, (4) rock, (5) bond twist, (6) ring formation, and (7) energy minimization (Lewis *et al.*, 1996).

A more recent study by Burton and Zones (Zones and Burton, 2005; Burton, 2007) used molecular simulations to determine the structure-directing influences of small amines in the crystallization of the MTT–zeolite phases. They predicted various novel amines and quaternary ammonium molecules with isopropyl, dimethylamino, tert-butyl, and trimethylamimonium groups connected via methylene spacers at 5 Å or multiples of 5 Å intervals that are able to crystallize the target MTT phases. The 5 Å spacers were found to be critical in the selective synthesis of specific phases. The molecules provide an optimal fit inside the MTT pores.

5.7 Summary and Future Directions

The close integration between theory and simulation with experiment has proven to be invaluable in establishing new insights into the structure and reactivity for a broad range of different catalytic materials and catalytic reactions. Theory has been instrumental in interpreting and resolving spectral data derived from infrared and Raman, ultraviolet, X-ray photoemission, near-edge absorption fine structure, nuclear magnetic resonance, scanning tunneling, temperature programmed desorption, and reaction studies. This synergistic combination of theory with spectroscopy has provided a wealth of information about the nature of the active site and the reactive intermediates for different catalytic systems. This detailed integration of theory and spectroscopy will undoubtedly continue and strengthen in the future with the advances in theory and characterization.

In addition to the advances in the prediction of structure and spectroscopy, theory and simulation have proven to be invaluable in the determination of complex potential energy surfaces and the corresponding kinetics for different surface processes. This has allowed for the elucidation of the elementary reaction steps, the influence of the reaction environment, and the resolution of rigorous catalytic kinetic studies. Future computational developments will expand the system sizes that can be simulated and thus significantly enhance our ability to elucidate the active site and the influence of its nanoscale environment (i.e., support, surface coverage, surface composition and structure, solution, applied potentials) on the intrinsic catalytic chemistry and kinetics, as well as on overall catalytic performance.

In just the past few years, theory and simulation have also emerged as valuable tools in establishing structure-reactivity relationships and aiding in the design of new materials. It is expected that both of these areas will increase significantly in the coming years. There will likely be more efforts focused not only on catalyst behavior but also on coarse-graining the surface simulations to detailed reactor models in an effort to link both catalyst design and optimal reactor performance.

In addition to the intrinsic catalytic activity and selectivity discussed, there are a number of other important practical issues that will likely drive both applications and the development of new methods in theory and simulations. These include catalyst

- Synthesis
- Stability

- Deactivation
- Durability.

All four of these are critical engineering considerations that will require the ability to simulate disparate changes in both length and time scales in order to bridge the atomic and molecular processes that govern these features to actual macroscopic behavior.

Theory and simulation have been used to examine a wide range of important applications, including fine chemical synthesis, NO_x reduction, selective oxidation, methanol and Fischer–Tropsch synthesis, chemo- and entioselective hydrogenation, electrocatalysis, alkylation, reforming, combustion, etc. It is very likely that most of the future applications will involve efforts in renewable energy: it is expected that there will be significant efforts in photocatalysis, electrocatalysis for fuel cells, and the conversion of biorenewable resources.

The significant advances in *ab initio* quantum mechanical methods, computational architectures, and computational power that have occurred over the past two decades are at the heart of the unprecedented advances in modeling catalytic systems. Despite these advances, improvements in method accuracy increase slowly. There are, however, major efforts in the theoretical chemistry and physics communities, which are aimed at the development of faster and more accurate methods to simulate complex materials. These include order-N scaling methods that can regularly simulate thousands of heavy atoms; more accurate DFT functionals, many body techniques, and quantum Monte Carlo methods that will offer significantly improved energies, band gaps, and electronic properties; improved parallelization schemes that will enable significant increases in system sizes; more robust transition-state search methods that will allow for rapid isolation of transition states; solid-state time-dependent DFT methods that will allow for a more accurate treatment of excited states and photocatalysis; and constant potential *ab initio* methods to simulate electrocatalytic systems.

The ability to simulate more complex issues such as catalyst synthesis, catalyst stability and durability, as well as photocatalysis and the conversion of renewable feedstocks will require the development of accurate coarse-grained models that enable the coupling between the electronic, atomic, nano, micro, and mesoscopic time and length scales. This will require further advances in robust force fields that can accurately treat reactive systems without heroic parametrization efforts, reactive molecular dynamics simulations, and "smart" kinetic Monte Carlo simulations that can accurately estimate internally within the simulation the activation barriers for the myriad of different reaction steps.

References

Allen, M. P., and Tildesley, D. J. (1987). *Computer Simulation of Liquids* (Clarendon Press, Oxford).

Andersson, M. P., Bligaard, T., Kustov, A., Larsen, K. E., Greeley, J., Johannessen, T., Christensen, C. H., and Nørskov, J. K. (2006). Toward computational screening in heterogeneous catalysis: Pareto-optimal methanation catalysts, *J. Catal.*, 239(2), pp. 501–506.

Arrouvel, C., Breysse, M., Toulhoat, H., and Raybaud, P. (2005). A density functional theory comparison of anatase (TiO_2)- and gamma-Al_2O_3-supported MoS_2 catalysts, *J. Catal.*, 232(1), pp. 161–178.

Barrett, P. A., Jones, R. H., Thomas, J. M., Sankar, G., Shannon, I. J., and Catlow, C. R. A. (1996). Rational design of a solid acid catalyst for the conversion of methanol to light alkenes: Synthesis, structure and performance of DAF-4, *Chem. Commun.*, 17, pp. 2001–2002.

Becke, A. D. (1985). Density-functional exchange-energy approximation with correct asymptotic behavior, *Phys. Rev. A*, 38, pp. 3098–3100.

Bell, A. T., Maginn, E. J., and Theodorou, D. N. (1997). Molecular simulation of adsorption and diffusion in zeolites, in *Handbook of Heterogeneous Catalysis*, eds. Ertl, H. K. G., and Weitkamp, J. (VCH, Weinheim), pp. 1165–1188.

Bligaard, T., Nørskov, J. K., Dahl, S., Matthiesen, J., Christensen, C. H., and Sehested, J. (2004). The Brønsted-Evans-Polanyi relation and the volcano curve in heterogeneous catalysis, *J. Catal.*, 224(1), pp. 206–217.

Boisen, A., Dahl, S., Nørskov, J. K., and Christensen, C. H. (2005). Why the optimal ammonia synthesis catalyst is not the optimal ammonia decomposition catalyst, *J. Catal.*, 230(2), pp. 309–312.

Burton, A. W. (2007). A priori phase prediction of zeolites: Case study of the structure-directing effects in the synthesis of MTT-type zeolites, *J. Am. Chem. Soc.*, 129(24), pp. 7627–7637.

Car, R., and Parrinello, M. (1985). Unified approach for molecular-dynamics and density-functional theory, *Phys. Rev. Lett.*, 55(22), pp. 2471–2474.

Carter, E. (2008). Challenges in modeling materials properties without experimental input, *Science*, 321, pp. 800–803.

Catlow, C. R. A., Coombes, D. S., and Pereira, J. C. G. (1998). Computer modeling of nucleation, growth, and templating in hydrothermal synthesis, *Chem. Mater.*, 10(11), pp. 3249–3265.

Costa, D., Arrouvel, C., Breysse, M., Toulhoat, H., and Raybaud, P. (2007). Edge wetting effects of gamma-Al_2O_3 and anatase-TiO_2 supports by MoS_2 and CoMoS active phases: A DFT study, *J. Catal.*, 246(2), pp. 325–343.

Csonka, G. I., Vydrov, O. A., Scuseria, G. E., Ruzsinszky, A., and Perdew, J. P. (2007). Diminished gradient dependence of density functionals: Constraint satisfaction and self-interaction correction, *J. Chem. Phys.*, 126(24), p. 244107.

Cummings, P. T. (2002). Molecular simulations and mesoscale methods, in *Applying Molecular and Materials Modeling*, eds. Westmoreland, P. R., Kollman, P. A., Chaka, A. M. *et al.* (Kluwer Academic Publishers, Dordrecht).

Davis, M., and Tilley, D. (2003). *NSF Workshop Report on Future Directions in Catalysis: Structures That Function on the Nanoscale* (National Science Foundation, Arlington, VA).

Delley, B. (2000a). DMol3 DFT studies: From molecules and molecular environments to surfaces and solids, *Comput. Mater. Sci.*, 17, pp. 122–126.

Delly, B. (2000b). From molecules to solids with the DMol3 approach, *J. Chem. Phys.*, 113, pp. 7756–7764.

Dumesic, J. A., Rudd, D. F., Aparicio, L. M., Rekoske, J. E., and Trevino, A. A. (1993). *The Microkinetics of Heterogeneous Catalysis* (American Chemical Society, Washington, DC).

Eichler, U., Kolmel, C. M., and Sauer, J. (1997). A computational study of the translational motion of protons in zeolite H-ZSM-5, *J. Comput. Chem.*, 18, p. 463.

Forseman, J. B., and Frisch, A. E. (1996). *Exploring Chemistry with Electronic Structure* (Gaussian, Inc., Pittsburgh, PA).

Frenkel, D., and Smit, B. (1996). *Understanding Molecular Simulation* (Academic Press Limited, San Diego, CA).

Froese, R. D. J., and Morokuma, K. (1999). Accurate calculations of bond-breaking energies in C60 using the three-layered ONIOM method, *Chem. Phys. Lett.*, 305, pp. 419–424.

Gale, J. D. (1997). GULP: A computer program for the symmetry adapted simulation of solids, *JCS Faraday Trans.*, 93, p. 629.

Gale, J. D and Rohl, A. L. (2003). The General Utility Lattice Program, *Mol. Simulat.*, 29, p. 291.

Giordano, L., Pacchioni, G., Goniakowski, J., Nilius, N., Rienks, E. D. L., and Freund, H. J. (2008). Charging of metal adatoms on ultrathin oxide films: Au and pd on FeO/Pt(111), *Phys. Rev. Lett.*, 101(2), p. 026102.

Goodman, D. W., Peden, C. H. F., and Chen, M. S. (2007). CO oxidation on ruthenium: The nature of the active catalytic surface, *Surf. Sci.*, 601(19), pp. L124–L126.

Govind, N., Wang, Y. A., da Silva, A. J. R., and Carter, E. A. (1999). Accurate ab initio energetics of extended systems via explicit correlation embedded in a density functional environment, *Chem. Phys. Lett.*, 295, pp. 129–134.

Hafner, J. (2008). Ab-initio simulations of materials using VASP: Density-functional theory and beyond, *J. Comput. Chem.*, 29(13), pp. 2044–2078.

Haubrich, J., Loffreda, D., Delbecq, F., Jugnet, Y., Sautet, P., Krupski, A., Becker, C., and Wandelt, K. (2006). Determination of the crotonaldehyde structures on Pt and PtSn surface alloys from a combined experimental and theoretical study, *Chem. Phys. Lett.*, 433(1–3), pp. 188–192.

Haubrich, J., Loffreda, D., Delbecq, F., Jugnet, Y., Sautet, P., Krupski, A., Becker, C., and Wandelt, K. (2008). Adsorption and vibrations of alpha, beta-unsaturated aldehydes on pure Pt and Pt-Sn alloy (111) surfaces I. Prenal, *J. Phys. Chem. C*, 112(10), pp. 3701–3718.

Head-Gordon, M. (1996). Quantum chemistry and molecular processes, *J. Phys. Chem.*, 100, p. 13213.

Hedin, L. (1965). New method for calculating the one-particle Green's function with application to the electron-gas problem, *Phys. Rev.*, 139, pp. A796–A823.

Hedin, L., and Lundqvist, S. (1969). In *Solid State Physics: Advances in Research and Application* 23:1, eds. Seitz, F., Turnbull, D., and Ehrenreich, H. (Academic Press, New York).

Hellman, A., Baerends, E. J., Biczysko, M., Bligaard, T., Christensen, C. H., Clary, D. C., Dahl, S., van Harrevelt, R., Honkala, K., Jonsson, H., Kroes, G. J., Luppi, M., Manthe, U., Nørskov, J. K., Olsen, R. A., Rossmeisl, J., Skulason, E., Tautermann, C. S., Varandas, A. J. C., and Vincent, J. K. (2006a). Predicting catalysis: Understanding ammonia synthesis from first-principles calculations, *J. Phys. Chem. B*, 110(36), pp. 17719–17735.

Hellman, A., Honkala, K., Remediakis, I. N., Logadottir, A., Carlsson, A., Dahl, S., Christensen, C. H., and Nørskov, J. K. (2006b). Insights into ammonia synthesis from first-principles, *Surf. Sci.*, 600(18), pp. 4264–4268.

Hermann, K., Witko, M., Druzinic, R., and Tokarz, R. (2001). Oxygen vacancies at oxide surfaces: Ab initio density functional theory studies on vanadium pentoxide, *App. Phys. A Mat. Sci. & Proc.*, 72(4), pp. 429–442.

Heyd, J., and Scuseria, G. E. (2003). Hybrid functionals based on a screened Coulomb potential, *J. Chem. Phys.*, 118, p. 8207.

Hohenberg, P., and Kohn, W. (1964). Inhomogeneous electron gas, *Phys. Rev.*, 136, p. B864.

Honkala, K., Hellman, A., Remediakis, I. N., Logadottir, A., Carlsson, A., Dahl, S., Christensen, C. H., and Nørskov, J. K. (2005). Ammonia synthesis from first-principles calculations, *Science*, 307(5709), pp. 55–558.

Hybertsen, M. S., and Louie, S. G. (1986). Spin-orbit splitting in semiconductors and insulators from the ab initio pseudopotential, *Phys. Rev. B*, 34, p. 5390.

Jacobsen, C. J. H., Dahl, S., Clausen, B. S., Bahn, S., Logadottir, A., and Nørskov, J. K. (2001). Catalyst design by interpolation in the periodic table: Bimetallic ammonia synthesis catalysts, *J. Am. Chem. Soc.*, 123(34), pp. 8404–8405.

Janesko, B. G., and Scuseria, G. E. (2007). Local hybrid functionals based on density matrix products, *J. Chem. Phys.*, 127(16).

Jensen, F. (1999). *Introduction to Computational Chemistry*, 2nd Ed. (Wiley, New York).

Kohn, W., and Sham, L. (1965). Self-consistent equations including exchange and correlation effects, *Phys. Rev.*, 140, p. A1133.

Kolczewski, C., and Hermann, K. (2003). Identification of oxygen sites at the $V_2O_5(010)$ surface by core-level electron spectroscopy: Ab initio cluster studies, *J. Chem. Phys.*, 118(16), pp. 7599–7609.

Kolczewski, C., and Hermann, K. (2004). Ab initio DFT cluster studies of angle-resolved NEXAFS spectra for differently coordinated oxygen at the V(2)O(5)(010) surface, *Surf. Sci.*, 552(1–3), pp. 98–110.

Kresse, G., and Furthmüller, J. (1996a). Efficiency of ab-initio total energy calculations for metals and semiconductors using a plane-wave basis set, *Comput. Mater. Sci.*, 6, pp. 15–50.

Kresse, G., and Furthmüller, J. (1996b). Efficient iterative schemes for ab-initio total energy calculations using a plane-wave basis set, *Phys. Rev. B*, 54, p. 11169.

Kresse, G., and Hafner, J. (1993). Ab initio molecular dynamics for liquid metals, *J. Phys. Rev. B*, 47, pp. 558–561.

Kustov, A. L., Frey, A. M., Larsen, K. E., Johannessen, T., Nørskov, J. K., and Christensen, C. H. (2007). CO methanation over supported bimetallic Ni-Fe catalysts: From computational studies towards catalyst optimization, *App. Catal. A: General*, 320, pp. 98–104.

Leach, A. R. (1996). *Molecular Modeling: Principles and Applications* (Pearson Education Limited, Essex).

Lewis, D. W., Catlow, C. R. A., and Thomas, J. M. (1997a). Application of computer modelling to the mechanisms of synthesis of microporous catalytic materials, *Farad. Disc.*, pp. 451–471.

Lewis, D. W., Freeman, C. M., and Catlow, C. R. A. (1995). Predicting the templating ability of organic additives for the synthesis of microporous materials, *J. Phys. Chem.*, 99(28), pp. 11194–11202.

Lewis, D. W., Sankar, G., Wyles, J. K., Thomas, J. M., Catlow, C. R. A., and Willock, D. J. (1997b). Synthesis of a small-pore microporous material using a computationally designed template, *Ange. Chem., Int. Ed.*, 36(23), pp. 2675–2677.

Lewis, D. W., Willock, D. J., Catlow, C. R. A., Thomas, J. M., and Hutchings, G. J. (1996). De novo design of structure-directing agents for the synthesis of microporous solids, *Nature*, 382(6592), pp. 604–606.

Logadottir, A., and Nørskov, J. K. (2003). Ammonia synthesis over a Ru(0001) surface studied by density functional calculations, *J. Catal.*, 220, pp. 273–279.

Malek, K., Li, C., and van Santen, R. A. (2007). New theoretical insights into epoxidation of alkenes by immobilized Mn-salen complexes in mesopores: Effects of substrate, linker and confinement, *J. Mol. Catal. A, Chem.*, 271(1–2), pp. 98–104.

Perdew, J. P., Chevary, J. A., Vosko, S. H., Jackson, K. A., Pederson, M. R., Singh, D. J., and Fiolhais, C. (1992). *Phys. Rev. B*, 46, p. 6671.

Perdew, J. P., Ruzsinszky, A., Tao, J. M., Staroverov, V. N., Scuseria, G. E., and Csonka, G. I. (2005). Prescription for the design and selection of density functional approximations: More constraint satisfaction with fewer fits, *J. Chem. Phys.*, 123(6).

Perdew, J. P., Tao, J. M., Staroverov, V. N., and Scuseria, G. E. (2004). Meta-generalized gradient approximation: Explanation of a realistic nonempirical density functional, *J. Chem. Phys.*, 120(15), pp. 6898–6911.

Philipsen, P. H. T., van Lenthe, E., Snijders, J. G., and Baerends, E. J. (1997). Relativistic calculations on the adsorption of CO on the (111) surfaces of Ni, Pd, and Pt within the zeroth-order regular approximation, *Phys. Rev. B*, 56, pp. 13556–13562.

Pisani, C., and Dovesi, R. (1980). Exact exchange Hartree–Fock calculations for periodic systems. I. Illustration of the method, *Int. J. Quan. Chem.*, 17, p. 501.

Pisani, C., Dovesi, R., and Roetti, C. (1988). *Hartree-Fock Ab-Initio Treatment of Crystalline Systems, Lecture Notes in Chemistry* (Springer-Verlag, Berlin and New York).

Raybaud, P. (2007). Understanding and predicting improved sulfide catalysts: Insights from first principles modeling, *App. Catal. A, General*, 322, pp. 76–91.

Raybaud, P., Costa, D., Valero, M. C., Arrouvel, C., Digne, M., Sautet, P., and Toulhoat, H. (2008). First principles surface thermodynamics of industrial supported catalysts in working conditions, *J. Phys. Cond. Matt.*, 20(6).

Reuter, K., Frenkel, D., and Scheffler, M. (2004). The steady state of heterogeneous catalysis, studied by first-principles statistical mechanics, *Phys. Rev. Lett.*, 93(11).

Reuter, K., and Scheffler, M. (2002). Composition, structure, and stability of $RuO_2(110)$ as a function of oxygen pressure, *Phys. Rev. B*, 65(3).

Reuter, K., and Scheffler, M. (2003a). Composition and structure of the $RuO_2(110)$ surface in an O-2 and CO environment: Implications for the catalytic formation of CO_2, *Phys. Rev. B*, 68(4).

Reuter, K., and Scheffler, M. (2003b). First-principles atomistic thermodynamics for oxidation catalysis: Surface phase diagrams and catalytically interesting regions, *Phys. Rev. Lett.*, 90(4).

Reuter, K., and Scheffler, M. (2006). First-principles kinetic Monte Carlo simulations for heterogeneous catalysis: Application to the CO oxidation at $RuO_2(110)$. *Phys. Rev. B*, 73(4).

Sauer, J. (1994). Structure and reactivity of zeolite catalysts: Atomistic modelling using *ab initio* techniques, *Stud. Surf. Sci. Catal.*, 84, pp. 2039–2057.

Sauer, J., and Sierka, M. (2000). Combining quantum mechanics and interatomic potential functions in *ab initio* studies of extended systems, *J. Comput. Chem.*, 21, pp. 1470–1493.

Saunders, V. R. (1984). Ab initio Hartree–Fock calculations for periodic systems, *Faraday Symp. Chem. Soc.*, 19, pp. 79–84.

Scheffler, M. (1988). *Physics of Solid Surfaces*, ed. Koukal, J. (Elsevier, Amsterdam).

Segall, M. D., Lindan, P. L. D., Probert, M. J., Pickard, C. J., Hasnip, P. J., Clark, S. J., and Payne, M. C. (2002). *J. Phys. Cond. Matt.*, 14(11), pp. 2717–2743.

Serrano-Andres, L., and Merchan, M. (2005). Quantum chemistry of the excited state: 2005 Overview, *J. Mol. Struct.*, 729, pp. 99–108.

Stampfl, C., Ganduglia-Pirovano, M. V., Reuter, K., and Scheffler, M. (2002). Catalysis and corrosion: The theoretical surface-science context, *Surf. Sci.*, 500(1–3), pp. 368–394.

Staroverov, V. N., Scuseria, G. E., Tao. J. M., and Perdew, J. P. (2004). Comparative assessment of a new nonempirical density functional: Molecules and hydrogen-bonded complexes, *J. Chem. Phys.*, 121(22), p. 11507.

Sterrer, M., Risse, T., Pozzoni, U. M., Giordano, L., Heyde, M., Rust, H. P., Pacchioni, G., and Freund, H. J. (2007). Control of the charge state of metal atoms on thin MgO films, *Phys. Rev. Lett.*, 98(9).

Sterrer, M., Yulikov, M., Fischbach, E., Heyde, M., Rust, H. P., Pacchioni, G., Risse, T., and Freund, H. J. (2006). Interaction of gold clusters with color centers on MgO(001) films, *Ange. Chem. Int. Ed.*, 45, pp. 2630–2632.

Studt, F., Abild-Pedersen, F., Bligaard, T., Sorensen, R. Z., Christensen, C. H., and Nørskov, J. K. (2008). Identification of non-precious metal alloy catalysts for selective hydrogenation of acetylene, *Science*, 320(5881), pp. 1320–1322.

Tao, J., Perdew, J. P., Ruzsinszky, A., Scuseria, G. E., Csonka, G. I., and Staroverov, V. N. (2007). Meta-generalized gradient approximation: Non-empirical construction and performance of a density functional, *Phil. Mag.*, 87(7), pp. 1071–1084.

Tao, J., Perdew, J. P., Staroverov, V. N., and Scuseria, G. E. (2003). Climbing the density functional ladder: Non empirical meta generalized gradient approximation designed for molecules and solids, *Phys. Rev. Lett.*, 91, p. 146401.

Tao, J. M., Staroverov, V. N., Scuseria, G. E., and Perdew, J. P. (2008). Exact-exchange energy density in the gauge of a semilocal density-functional approximation, *Phys. Rev. A*, 77(1).

Ulrich, S., Nilius, N., Freund, H. J., Martinez, U., Giordano, L., and Pacchioni, G. (2008). Evidence for a size-selective adsorption mechanism on oxide surfaces: Pd and Au atoms on SiO2/Mo(112), *ChemPhysChem*, 9(10), pp. 1367–1370.

US Department of Energy (2007). Basic research needs: Catalysis for energy. Report from the Basic Energy Sciences Workshop, DOE, Washington, DC 6–8 August 2007, http://www.sc.doe.gov/bes/reports/list.html.

van Santen, R. A., and Neurock, M. (2006). *Molecular Heterogeneous Catalysis: A Conceptual and Computational Approach* (Wiley-VCH, Weinheim).

Velde, G. te, Bickelhaupt, F. M., Baerends, E. J., Guerra, C. F., van Gisbergen, S. J. A., Snijders, J. G., and Ziegler, T. (2001). Chemistry with ADF, *J. Comput. Chem.*, 22(9), pp. 931–967.

Vreven, T., and Morokuma, K. (2000). *J. Chem. Phys.*, 113, pp. 2969–2975.

Westmoreland, P., Kollman, P. A., Chaka, A. M., Cummings, P. T., Morokuma, K., Neurock, M., Stechel, E. B., and Vashishta, P. (2002). *Applying Molecular and Materials Modeling* (Kluwer Academic Publishers, Dordrecht and Boston).

Whitten, J., and Yang, H. (1996). Theory of chemisorption and reactions on metal surfaces, *Surf. Sci. Rep.*, 24, p. 55.

Willock, D. J., Lewis, D. W., Catlow, C. R. A., Hutchings, G. J., and Thomas, J. M. (1997). Designing templates for the synthesis of microporous solids using *de novo* molecular design methods, *J. Mol. Catal. A, Chem.*, 119(1–3), pp. 415–424.

Yulikov, M., Sterrer, M., Heyde, M., Rust, H. P., Risse, T., Freund, H. J., Pacchioni, G., and Scagnelli, A. (2006). Binding of single gold atoms on thin MgO(001) films, *Phys. Rev. Lett.*, 96(14).

Zhang, H. D., Wang, Y. M., Zhang, L., Gerritsen, G., Abbenhuis, H. C. L., van Santen, R. A., and Li, C. (2008). Enantioselective epoxidation of beta-methylstyrene catalyzed by immobilized Mn(salen) catalysts in different mesoporous silica supports, *J. Catal.*, 256(2), pp. 226–236.

Zones, S. I., and Burton, A. W. (2005). Diquaternary structure-directing agents built upon charged imidazolium ring centers and their use in synthesis of one-dimensional pore zeolites, *J. Mater. Chem.*, 15(39), pp. 4215–4223.

6

Applications: Energy from Fossil Resources

Levi Thompson

6.1 Introduction

Fossil resources, including petroleum, natural gas, and coal, presently account for nearly 85% of all the energy consumed in the world (EIA, 2008; Figure 6.1) and the emission of nearly 30 billion tons of CO_2 annually (IPCC, 2007). The United States accounts for approximately 20% both of total energy resources consumed and CO_2 emitted (Figure 6.2) worldwide. With energy consumption predicted to double in the next 50 years and fossil fuels expected to satisfy much of this demand, there is growing interest in improved processes and materials to convert these fossil resources into fuels and remediate the CO_2 and other greenhouse gases emitted during combustion. Catalysts and catalytic processes will play key roles in both areas.

This chapter examines research and development activities related to the conversion of fossil resources, with an emphasis on catalysis for the production of liquid transportation fuels and hydrogen, and for emission control including the reduction of CO_2. Efforts in these areas span significant length and size scales, with work in Asia focusing on practical

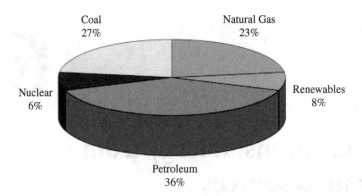

Fig. 6.1. According to the US Energy Information Administration, 482.5 quadrillion BTUs of energy were consumed globally in 2007, with more than 86% coming from fossil resources. This represents an increase of 21% from the 2000 consumption rate (statistics, EIA, 2008).

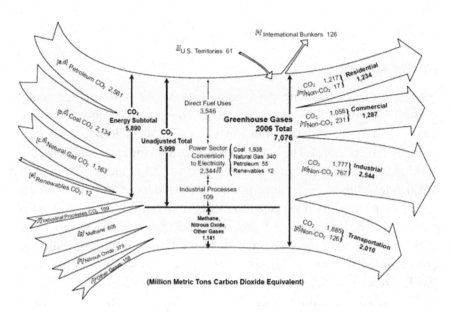

Fig. 6.2. Carbon dioxide emission flow chart illustrating amounts and sources for the United States in 2006 (EIA, 2007).

catalysts and processes with large-scale demonstration efforts, and work in Europe focusing on more fundamental investigations of catalyst function and the discovery of new formulations and processes. The following sections provide snapshots of information shared with the WTEC delegation of US engineers and scientists during visits to key institutions in Asia and Europe. In some cases findings from the site visits are supplemented with information from the literature.

6.2 Production of Liquid Fuels

6.2.1 Catalysts for petroleum refining

Nearly 95% of all liquid transportation fuels are produced from petroleum. The refining of petroleum into gasoline, diesel, jet fuels, and other transportation fuels involves a fairly complex network of processes. The key catalytic processes are catalytic reforming, hydrotreating (e.g., hydrodesulfurization and hydrodenitrogenation), and catalytic cracking (Figure 6.3). Much of the catalyst research for petroleum refining observed and described during the WTEC panel's visits focused on sulfur removal. The method used to remove sulfur depends on whether sulfur is present as hydrogen sulfide (H_2S) or organosulfur compounds. Two methods have been developed to remove H_2S; adsorption and absorption. Zinc oxide (ZnO) is most frequently used to adsorb H_2S (Othmer, 1994; Armor, 1999). Zinc oxide reacts with H_2S at 340–390°C to form zinc sulfide, ZnS (Armor, 1999). Commonly used absorbents are aqueous solutions of organic amines such as monoethanol amine, diethanol amine, and triethanol amine.[1]

The removal of organosulfur compounds via catalytic hydrodesulfurization (HDS) is typically carried out using aluminum oxide (Al_2O_3)-supported cobalt-molybdenum (Co-Mo)- or nickel-molybdenum (Ni-Mo)-based catalysts. This reaction converts organosulfur compounds into H_2S, which can be subsequently removed. Although the operating temperature for hydrodesulfurization generally ranges from 300–400°C, the

---

[1] For additional information about hydrodesulfurization and other hydrotreating reactions, the reader is directed to books by Topsøe and Carnell (e.g., Topsøe et al., 1996; Carnell, 1989).

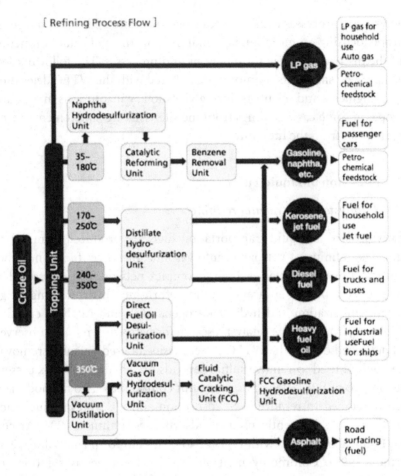

Fig. 6.3. Image of refinery unit and process flow diagram for the conversion of crude oil into low-sulfur commercial liquid and gas products (http://www.cosmo-oil.co.jp/eng/oilbusiness/refining.html).

operating pressure and the space velocity depend on the nature of the sulfur compounds to be removed. The severity of conditions typically increases in the following order: mercaptan < benzothiophene < alkyl benzothiophene < dibenzothiophene < alkyl dibenzothiophene. Organosulfur compounds in natural gas can be removed using low pressures; however, significantly higher pressures (20–40 atm) are typically required to desulfurize liquid fuels including naphtha. Deep hydrodesulfurization typically involves

Applications: Energy from Fossil Resources

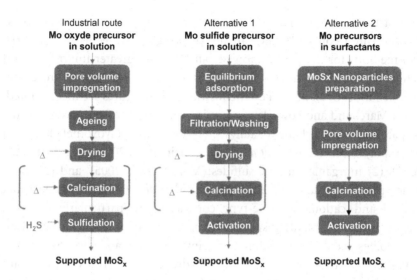

Fig. 6.4. Effects of precursor and preparation conditions on structure and function of hydrotreating catalysts: typical; (*left*) 2 alternatives being studied by IFP (*center and right*) (courtesy of K. Marchand, IFP).

pressures in the range of 60 to 100 atm and relatively low liquid hourly space velocities (LHSV) of 0.75 to 1.0 hr^{-1} to remove the most refractory sulfur compounds (e.g., 4- or 6-methyl dibenzothiophene and 4,6-dimethyl dibenzothiophene). Improved materials and processes are being sought to reduce the consumption of hydrogen and improve product selectivities.

As with most catalysts, performance is determined by the composition and structure, which are in turn determined by the preparation methods. Conventional methods for the preparation of hydrotreating catalysts initially proceed through oxide precursors supported on high-surface-area carriers (Figure 6.4). These oxide domains are sulfided to produce catalytic species that typically are nonuniform in terms of chemistry and morphology.

6.2.1.1 *Institut Français du Pétrole*

Researchers at the Institut Français du Pétrole (IFP) (see Site Reports-Europe on the International Assessment of Research in Catalysis by

Nanostructured Materials website, www.wtec.org/private/catalysis; password required, contact WTEC) are investigating novel precursors, supports, and methods for the preparation of catalysts with greater uniformity and better performance. One strategy involves the use of well-defined precursors in which most or all of the key chemistries and structures are incorporated. Karin Marchand and coworkers are using three types of precursors for the preparation of well-defined catalysts: (1) inorganic polyoxometallates and heteropolyanions (Cabello *et al.*, 2000; Martin *et al.*, 2004; Bergwerff *et al.*, 2005), (2) inorganic cationic sulfides (Martin *et al.*, 2005), and (3) MoS_x nanoparticles (Costa *et al.*, 2007; Marchand *et al.*, 2002). They report that neutral and cationic cubane-type complexes are particularly effective, yielding uniform, highly dispersed domains on supports including Al_2O_3 and zeolites and catalysts with high intrinsic activities and good selectivities to C–S bond activation over C=C bond hydrogenation. This is an important finding because the interaction of most organometallic clusters with oxide supports causes decomposition and rearrangement of the elements (Choplin *et al.*, 1993; Brenner and Thompson, 1994; Gates, 1995; 2000; Chotisuwan *et al.*, 2007).

The IFP researchers also reported that relatively narrow particle sizes were achieved for MoS_x nanoparticles synthesized using a reverse microemulsion technique, although inhibition of HDS by residue from the surfactants could be a challenge. Raybaud and coworkers are using density functional theory (DFT) to model key interactions of the metal and metal sulfide moieties with oxide supports, including Al_2O_3 and TiO_2, with a focus on defining the density of active sites (Digne *et al.*, 2002; 2004; Arrouvel *et al.*, 2004b; Dzwigaj *et al.*, 2005; Arrouvel *et al.*, 2004a; Raybaud, 2007; Arrouvel *et al.*, 2006). A particularly interesting observation was that MoS_2 can grow "epitaxially" on TiO_2, thus producing a high degree of M-edge sites (Figure 6.5) that are believed to be hosts for the catalytically active sites for HDS (Schweiger *et al.*, 2002).

6.2.1.2 *Consiglio Nazionale delle Ricerche (Italy)*

Francesca Deganello and coworkers at the Consiglio Nazionale delle Ricerche (CNR) Institute of Nanostructured Materials (INSM) in Palermo, Italy (see Site Reports-Europe on the International Assessment

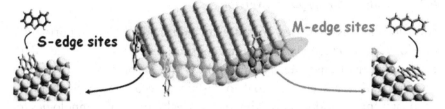

Fig. 6.5. Schematic of Mo sulfide domain on oxide support with coordination of organosulfur and aromatic species (Raybaud, 2007).

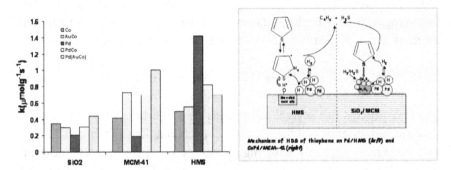

Fig. 6.6. Mesoporous silica-supported Co, Au, and/or Pd catalysts for thiophene HDS. The superior activity, in particular for the Pd/HMS catalyst, is explained based on a bifunctional mechanism involving acid sites on the support (CNR/ISMN; courtesy of R. Psaro).

of Research in Catalysis by Nanostructured Materials website, www.wtec.org/private/catalysis; password required, contact WTEC) are investigating the use of mesoporous oxides including hexagonal mesoporous silica (HMS) as "active" supports for cobalt, gold, and lead. As illustrated in Figure 6.6, the resulting catalysts are active and selective for thiophene HDS, and a bifunctional mechanism was proposed with Brønsted acid sites serving to activate the organosulfur compound for protonation on the metal domains (Venezia et al., 2007a; 2007b; Liotta et al., 2007a; 2007b). This theme of deconvoluting the reaction pathways and producing domains to catalyze the key intermediate steps provided guidance for the design of catalysts for HDS and other reactions.

6.2.2 Catalysts for syngas conversion

Globally there is increased pressure to reduce fossil fuel imports and develop technologies to convert indigenous resources into liquid fuels that can be used and transported conveniently. With 13% of the world's proven reserves, China has enough coal to sustain its economic growth for a century or more, and it is investing in the development of technologies to convert coal into liquid transportation fuels. An attractive strategy involves the oxidation of coal into mixtures of CO and H_2 (syngas), then using a Fischer–Tropsch synthesis (FTS) process, converting this syngas into liquid fuels. The FTS reaction produces hydrocarbons and oxygenates from syngas that can be derived from any carbonaceous material, including biomass and coal (Bartholomew, 1990; DOE, 2008; Equation (6.1)).

$$CO + 2H_2 \rightarrow \text{-}CH_2\text{-} + H_2O. \quad \Delta H° = -165 \text{ kJ/mol} \quad (6.1)$$

This process has several advantages over the biochemical processes that are being developed for biomass conversion. The FTS hydrocarbons and oxygenates typically have higher energy densities than ethanol and biodiesel and can be blended with or substituted directly for conventional liquid transportation fuels such as diesel and gasoline. In addition, syngas can be produced from a wide range of carbonaceous materials, including the cellulose-rich portion of biomass (e.g., stover) that is difficult to convert using biochemical processes. Relative to conventional diesel, synthetic liquid fuels from FTS produce significantly lower NO_x, particulate, hydrocarbon, and CO emissions (Kochloefl, 1997). The water gas shift (WGS) reaction is typically used to increase the H_2/CO ratios for syngas to levels suitable for FTS, and the FTS is carried out at 180–250°C and pressures ranging from 20–40 bar using Co-, Fe-, and Ni-based catalysts. Improvements in selectivity and activity are being sought through refinements in the catalyst structure and composition.

6.2.2.1 *University of Utrecht (The Netherlands)*

Krijn de Jong from Utrecht University (see Site Reports-Europe on the International Assessment of Research in Catalysis by Nanostructured

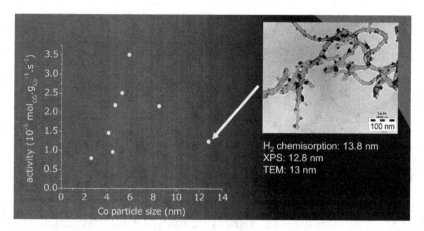

Fig. 6.7. Effect of particle size on Fischer–Tropsch synthesis activities of carbon-nanofiber-supported Co catalysts. Measurements were carried out at 220°C, 1 bar, and a H_2/CO ratio of 2 (Bezemer, 2006).

Materials website, www.wtec.org/private/catalysis; password required, contact WTEC) described to the WTEC panel work to determine the effect of particle size on FTS activity. In one example, the FTS activity for a Co/C catalyst was illustrated to go through a maximum for Co particles that were ~6 nm (Bezemer, 2006; Figure 6.7). The chemistry and structure of these catalysts were characterized using chemisorption, X-ray photoelectron spectroscopy (XPS) and transmission electron microscopy (TEM).

This group also investigated the effects of different calcination gases and conditions on the morphological and functional properties of oxide-supported Co catalysts. For example, they have reported that Co/SiO_2 catalysts calcined in NO possessed much smaller Co particles than catalysts calcined at the same temperature in air (Figure 6.8). As expected, reaction rates for the NO-calcined materials were higher than those for the air-calcined materials.

6.2.2.2 Institute of Coal Chemistry/Chinese Academy of Sciences (China)

Researchers at the Institute of Coal Chemistry of the Chinese Academy of Sciences (ICC/CAS; see Site Reports-Asia on the International Assessment of Research in Catalysis by Nanostructured Materials website, www.wtec.

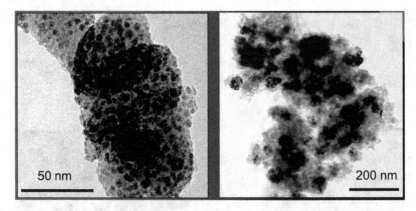

Fig. 6.8. Transmission electron micrographs of Co/SiO$_2$ catalysts following calcination in (a) NO and (b) air, then reduction at 550°C. Particle sizes were significantly smaller for the NO-calcined catalyst, and the Fischer–Tropsch synthesis reaction rates were higher (courtesy of K. De Jong).

org/private/catalysis; password required, contact WTEC) are developing nanostructured catalysts for the FTS and CH$_4$ (methane) dry reforming reactions using methods ranging from quantum chemistry and molecular dynamics to reactor simulation. The efforts have yielded new combinations of catalysts, including Fe carbides in MoS$_2$, and a potentially new nanophase MoS$_2$ was described by Yuhan Sun and coworkers (Li *et al.*, 2004; 2007). The performance of a variety of catalysts is being evaluated in major pilot plant facilities. Figure 6.9 illustrates pilot plants for coal gasification and FTS. These units have the capacity to convert 100 tons of coal per day into syngas and then Fischer–Tropsch liquids. Work at this scale is typically not found in academic or government laboratories in other countries; however, investments of this scale are increasing in China.

6.3 Production of Hydrogen

More than 50 million tons of hydrogen are produced globally each year. Nearly half of this hydrogen is derived from natural gas, 30% from petroleum, ~18% by reducing water with coal, and ~2% from water via electrolysis. The conversion of fossil resources into hydrogen typically involves a desulfurizer to remove sulfur compounds from the feed,

Fig. 6.9. (*Left*) ICC/CAS can process 100 tons/day long rank coal in a fluidized bed gasification pilot/demonstration plant; a higher-pressure process is under development for higher capacity, (*right*) a large Fischer–Tropsch slurry demonstration unit (Fe catalysts).

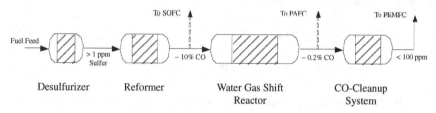

Fig. 6.10. Schematic of key steps in the conversion of fossil resources into hydrogen for fuel cell applications.

reformer to produce syngas, and a CO-cleanup step to reduce the amount of CO (Figure 6.10). Carbon monoxide that is in reformer effluent can severely and irreversibly poison downstream catalysts, including those used in ammonia synthesis and proton exchange membrane fuel cells (PEMFC); therefore CO must be reduced to ppm levels (Carrette *et al.*, 2000; Parsons and Vandernoot, 1988; Capon and Parsons, 1973). This is accomplished via the water gas shift (Equation (6.4)), followed by CO preferential oxidation or methanation.

The syngas is most often produced using steam reforming and/or partial oxidation reactions. Key reactions in the steam reformer are listed below.

$$CH_4 + H_2O \rightarrow CO + 3H_2, \quad \Delta H° = 206 \text{ kJ/mol} \qquad (6.2)$$

$$C_nH_{2n+2} + mH_2O \rightarrow nCO + (m+n+1)H_2, \tag{6.3}$$

$$CO + H_2O \rightarrow CO_2 + H_2. \quad \Delta H° = -41 \text{ kJ/mol} \tag{6.4}$$

The reforming of higher hydrocarbons typically starts with decomposition into lower-molecular-weight species that can more efficiently react with steam. Oxide-supported Fe, Co, and Ru catalysts are effective for the FTS reaction, although they can suffer from deactivation by coke deposition. This coke is produced through polymerization of the various reactive olefinic species on the surface. The partial oxidation reactions are exothermic and can be carried out without (noncatalytic partial oxidation) or with a catalyst (catalytic partial oxidation). The main advantage of the noncatalytic partial oxidation process is that it can be used to process almost any hydrocarbon feedstock from natural gas to petroleum residue (Equations (6.5) and (6.6)).

$$CH_4 + 0.5\, O_2 \rightarrow CO + 2H_2, \quad \Delta H° = -36 \text{ kJ/mol} \tag{6.5}$$

$$2C_nH_{2n+2} + nO_2 \rightarrow 2nCO + 2(n+1)H_2. \tag{6.6}$$

Other reactions like total combustion and steam reforming also take place to some extent during the partial oxidation process (Pena et al., 1996). Typically, noncatalytic partial oxidation reactions are carried out at very high temperatures and employ smaller reactors than those used for steam reforming reactions. Autothermal reforming combines partial oxidation and steam reforming to create a near thermally neutral process. The partial or total oxidation of the fuel is carried out in the front end of the reactor and is followed by steam reforming. The heat generated via partial oxidation provides the heat required for the endothermic steam reforming reactions. Suitable feedstocks for autothermal reforming are CH_4-rich natural gas, CH_3OH and heavier hydrocarbons, up through naphtha (Gunardson, 1998).

The water gas shift reactor typically constitutes a third of the size and cost of the systems used to convert fossil resources into hydrogen. As such, it is not surprising that much of the research in the area of hydrogen production focuses on the development of better-performing WGS catalysts. The WGS reaction is commercially carried out in two stages with a

Applications: Energy from Fossil Resources 163

high-temperature-shift (HTS) stage employing Fe-Cr-based catalysts and a low-temperature-shift (LTS) stage using Cu-Zn-Al-based formulations. While the Fe-Cr HTS catalysts are robust and to some extent sulfur-tolerant, their activities are low. Furthermore, some countries prohibit the use of these Fe-Cr-based catalysts because of the potential for producing genotoxic hexavalent Cr (EPA, 1998). The Cu-Zn-Al LTS catalysts are highly active but very susceptible to deactivation.

6.3.1 University of Udine, Consiglio Nazionale delle Ricerche (Italy)

Researchers at the University of Udine, in connection with the CNR (see CNR site report, Site Reports-Europe on the International Assessment of Research in Catalysis by Nanostructured Materials website, www.wtec.org/private/catalysis; password required, contact WTEC), are manipulating characteristics of the metal particles and supports to enhance the overall reaction rate for WGS. Alessandro Trovarelli and coworkers described tailor-made rare earth oxide- and ZrO_2-supported Au and Pt catalysts with high activities (Tibiletti et al., 2006). They also reported that the {100} surface of CeO_2, a support that is used in several high-activity WGS catalyst formulations, is more active for the reduction of water than the other facets (Aneggi et al., 2005). Methods are being developed to increase the exposure of {100} facets for CeO_2-supported catalysts. Figure 6.11 shows high-resolution transmission electron micrographs illustrating CeO_2 supports with high preferential exposures {100} planes.

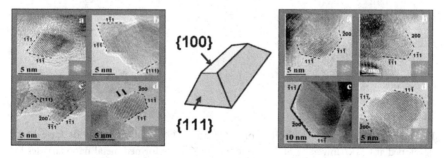

Fig. 6.11. CNR/University of Udine researchers are developing novel water gas shift catalysts via support optimization. For CeO_2 supported catalysts, supports are modified to enhance CO oxidation, preferentially exposing the more active {100} surface.

6.3.2 University of Trieste, Consiglio Nazionale delle Ricerche (Italy)

Researchers at Italy's University of Trieste, in connection with CNR (see Site Reports-Europe on the International Assessment of Research in Catalysis by Nanostructured Materials website, www.wtec.org/private/catalysis; password required, contact WTEC), are working on preferential oxidation (PrOx) and WGS catalysts for the purification of hydrogen-rich reformate, the design of nanostructured photocatalysts, and development of mixed-oxide cathodes for solid oxide fuel cells. Mauro Graziani and coworkers described methods to encapsulate preformed metal nanoparticles in oxides as a strategy to enhance catalyst durability (Montini et al., 2007a; Montini et al., 2007b; Figure 6.12). Active species, including Rh, Ni, and Cu, have been encapsulated in porous Al_2O_3 and CeO_2-ZrO_2-Al_2O_3. They argued that this architecture inhibits sintering of the active species but allows reactants and products to diffuse through pores in the oxide. Enhanced durability was demonstrated for several WGS and preferential oxidation catalysts, including Au/CeO_2.

6.3.3 Instituto di Chimica dei Composti OrganoMetallici (Italy)

Claudio Bianchini and coworkers from the Instituto di Chimica dei Composti OrganoMetallici (University of Florence and CNR) (see Site Reports-Europe on the International Assessment of Research in Catalysis by

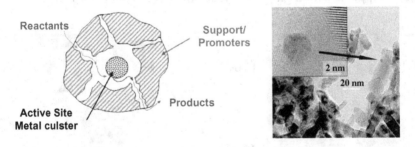

Fig. 6.12. To enhance the stability of PrOx and WGS catalysts, methods have been developed at CNR/University of Trieste to encapsulate preformed metal nanoparticles in porous oxides. This architecture (*left*) allows the reactants and products access to the catalytically active metal while inhibiting sintering. A high-resolution transmission electron micrograph (*right*) illustrating an encapsulated metal particle.

Nanostructured Materials website, www.wtec.org/private/catalysis; password required, contact WTEC) are investigating catalytic systems for the production of hydrogen, including Fe-Ni-Co/Al_2O_3 ethanol steam reforming catalysts, and electrocatalysts for the electrolysis of H_2O, NH_3, and renewable feedstocks such as glycerol. They are also detailing the reaction pathways for these systems. For example, they propose that the conversion of glycerol proceeds through several value-added intermediates, including glyceric acid, tartronic acid, glycolic acid, and oxalic acid, to form formic acid and CO_2 as products.

6.3.4 Tokyo Metropolitan University Department of Applied Chemistry

Haruta and coworkers at Tokyo Metropolitan University (TMU) Department of Applied Chemistry (see Site Reports-Asia on the International Assessment of Research in Catalysis by Nanostructured Materials website, www.wtec.org/private/catalysis; password required, contact WTEC) have done pioneering work in the area of nanostructured Au catalysts and continue to develop these materials for a variety of applications, including WGS (Haruta et al., 1987; Tsubota et al., 1991; Sakurai et al., 1993; Haruta et al., 1993; Haruta and Date, 2001; Haruta, 2002; Ishida and Haruta, 2007). Their work has yielded catalysts with nanoscale Au particles (Figure 6.13) that are more active than most other materials, including Cu-Zn-Al and supported Pt catalysts. They report 100% conversion to CO_2 with no CH_4 production at high space velocities and temperatures up to 623 K. Stability issues, however, remain to be addressed, because these materials tended to deactivate over time, in particular when exposed to high levels of moisture (Kolmakov and Goodman, 2001; Costello et al., 2003; Kim and Thompson, 2005; Goguet et al., 2007; Karpenko et al., 2007).

6.3.5 Tsinghua University (China)

Research at Tsinghua University (see Site Reports-Asia on the International Assessment of Research in Catalysis by Nanostructured Materials website, www.wtec.org/private/catalysis; password required, contact WTEC) is investigating ways to control the structure of catalysts.

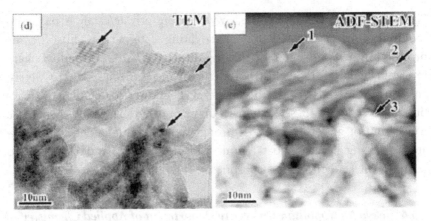

Fig. 6.13. High-resolution transmission electron micrographs of highly active Au/CeO$_2$ WGS catalysts. These materials achieved 100% conversion to CO$_2$ with no production of methane for temperatures up to 623 K.

Boqing Xu and coworkers describe the use of electrochemical methods to produce nanostructured materials. High dispersions of metals including Ni, Au, and Pt have been achieved for a variety of support materials including ZrO$_2$ and TiO$_2$ (Wang et al., 2005; Wu et al., 2005; Li et al., 2006; Wu et al., 2007a; Yu and Xu, 2006; 2007; Wu et al., 2007b; Xin et al., 2005a; Xin et al., 2005b; Zhang et al., 2007). The resulting catalysts were active for the dry reforming, CO oxidation, diene hydrogenation, and hydrogen electrocatalytic oxidation reactions.

6.3.6 Tianjin University (China)

Changjun Liu from Tianjin University (China) (see Site Reports-Asia on the International Assessment of Research in Catalysis by Nanostructured Materials website, www.wtec.org/private/catalysis; password required, contact WTEC) described the use of plasma treatment methods to alter the structure and function of catalytic materials. This low-temperature treatment yielded structures that are very different from those for catalysts produced using more conventional methods. In one example, they observed that the orientation of Pt particles on TiO$_2$ subject to plasma treatment was different from the orientation of those produced using thermal treatment (Figure 6.14). The plasma treatment process was also

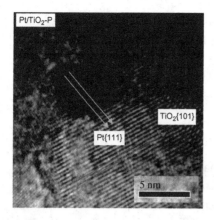 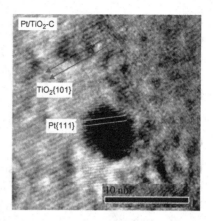

Fig. 6.14. High-resolution transmission electron micrographs comparing the orientations of Pt particles relative to the TiO$_2$ supports for catalysts treated using (*left*) conventional high-temperature thermal treatments and (*right*) low-temperature, plasma-based treatments.

used to produce Ni/Al$_2$O$_3$ CH$_4$ dry reforming catalysts with higher activities and reduced susceptibility to coking.

6.4 Fuel Cell Research

Key challenges for fuel cells include their high cost and poor durability. There is substantial research on electrocatalysts for fuel cells in both Asia and Europe; however, the WTEC panel was only able to visit a few laboratories doing fuel cell research. Several types of fuel cell are being investigated, including polymer electrolyte membrane fuel cells (PEMFC), solid oxide fuel cells (SOFC), as well as molten carbonate, direct methanol, and direct carbon fuel cells. Much of the work can be characterized as applied research and development.

6.4.1 *University of Trieste, Consiglio Nazionale delle Ricerche (Italy)*

Graziani and coworkers from University of Trieste (see Site Reports-Europe on the International Assessment of Research in Catalysis by Nanostructured Materials website, www.wtec.org/private/catalysis; password required, contact WTEC) are developing $LaNi_{0.6}Fe_{0.4}O_3$ and $La_{(1-y)}Sr_yNi_{0.6}Fe_{0.4}O_3$ cathodes for SOFCs. This material was reported to have much higher ionic conductivities than $LaNi_{0.6}Fe_{0.4}O_3$ (Bevilacqua et al., 2006). Experimental and theoretical studies are being used to optimize performance characteristics of these materials.

6.4.2 Tsinghua University (China)

Boqing Xu and coworkers from Tsinghua University (see Site Reports-Asia on the International Assessment of Research in Catalysis by Nanostructured Materials website, www.wtec.org/private/catalysis; password required, contact WTEC) are developing methods to disperse Pt on Au cores for use in PEMFCs (Zhao et al., 2005; Zhao and Xu, 2006). This architecture would allow greater utilization of the expensive Pt. These catalysts achieved near 100% Pt utilization due to the dispersion of Pt domains smaller than 1 nm. The synthetic process can also be used to create core-shell geometries with Pt shells on Au cores.

6.4.3 Tianjin University (China)

Yongdan Li from Tianjin University (see Site Reports-Asia on the International Assessment of Research in Catalysis by Nanostructured Materials website, www.wtec.org/private/catalysis; password required, contact WTEC) described a relatively novel type of fuel cell, the direct carbon fuel cell (DCFC). Carbon is electrochemically oxidized using graphite electrodes and molten carbonate electrolytes according to the following reactions (Equations (6.7) and (6.8)):

$$C + 4OH^- \rightarrow CO_2 + 2H_2O + 4e^-, \tag{6.7}$$

$$O_2 + 2H_2O + 4e^- \rightarrow 4OH^-. \tag{6.8}$$

Efficiencies near 40% can be achieved with this type of device. A hybrid concept was described with a reactor to decompose CH_4 to carbon, a DCFC to oxidize the carbon, a PEMFC to convert the H_2, and a waste heat recovery system (Cao et al., 2004; Chen et al., 2004; Qian et al., 2003a;

Qian et al., 2003b; He et al., 2006; Chen et al., 2003). Overall efficiencies near 100% were suggested. These systems could be used as an alternative to coal-fired power plants.

In addition to efforts to develop more durable and lower-cost catalysts, research at the *Dalian Institute of Chemical Physics* (see Site Reports-Asia on the International Assessment of Research in Catalysis by Nanostructured Materials website, www.wtec.org/private/catalysis; password required, contact WTEC) and *Shanghai Jiao Tong University* focuses on large-scale, multistack demonstrations. The associated laboratories are large and fairly well equipped.

6.5 Environmental Catalysis

Environmental catalysis remains an important area with regard to the conversion of fossil fuels. With the increasing number of vehicles, in particular in developing countries, the need for more effective emission control catalysts is growing. There is also growing interest in the development of processes to recycle CO_2 into useful products. Nearly 30 billion tons of CO_2 are emitted into the environment annually due to human activity, including transportation and electricity production. With growing evidence linking this CO_2 to disruptive global climate change and predictions that CO_2 emissions will double in the next 50 years, the remediation of CO_2 has attracted considerable attention. In a recent report from the National Academies, managing atmospheric CO_2 is listed as one of 14 Grand Challenges for the 21st Century — along with major scientific and engineering challenges like the discovery of better medicines, securing cyberspace, and providing access to clean water (National Academy of Engineering, 2008). Technologies are being developed to capture and sequester CO_2; however, there are concerns about the capacities and the long-term environmental impacts of storing CO_2 in, for example, geologic formations. Of course, biological processes including photosynthesis provide highly selective routes for CO_2 and H_2O conversion into carbohydrates, but the rates are too low for large-scale conversion of anthropogenic CO_2. Consequently, these are opportunities for the discovery and development of new catalysts.

6.6 Three-Way Catalysis

6.6.1 *Toyota Motor Corporation, Higashi-Fuji Technical Center*

Researchers at the Toyota Motor Corporation's Higashi-Fuji Technical Center (see Site Reports-Asia on the International Assessment of Research in Catalysis by Nanostructured Materials website, www.wtec.org/private/catalysis; password required, contact WTEC) are working to deconvolute the complex chemistry and behavior of three-way catalysts, in terms of both reactivity and deactivation. Results from high-resolution transmission electron microscopy, X-ray absorption spectroscopy, and X-ray photoelectron spectroscopy indicate that Pt on Al_2O_3-ceria-zirconia-yttria (CZY) support produces two different types of sites. Platinum that resides on the Al_2O_3 is weakly bound and agglomerates into zero-valent particles; platinum on the CZY is more strongly bound, maintains a high degree of dispersion, and is oxidized. Figure 6.15 illustrates the presence of relatively large Pt particles on Al_2O_3 portions of the catalyst and highly dispersed domains on the CZY portion of the catalyst.

6.6.2 *NO_x selective catalytic reduction*

6.6.2.1 *Toyota Motor Corporation, Higashi-Fuji Technical Center*

Researchers at Toyota Motor Corporation's Higashi-Fuji Technical Center are also working on NO_x storage materials, most often incorporating barium. A variety of tools are being used in this research to provide insights regarding sites where oxygen interacts with barium, the mechanism for oxygen migration to platinum sites, changes in the chemistry of the platinum site, and ultimately, mechanisms for the production of nitrates on the surface. Of particular importance is work in collaboration with Michael Bowker at Cardiff University (UK) (see Site Reports-Europe on the International Assessment of Research in Catalysis by Nanostructured Materials website, www.wtec.org/private/catalysis; password required, contact WTEC) using scanning tunneling microscopy (STM). The results suggest that NO_x storage proceeds via dissociation of O_2 on the Pt step edges, then reaction of atomic oxygen with barium on the terraces to produce BaO and BaO_2. Subsequently, NO and O_2 react with BaO_2 on the terrace to form O-Ba-NO_2 species. The adsorbed NO_2 can diffuse back to

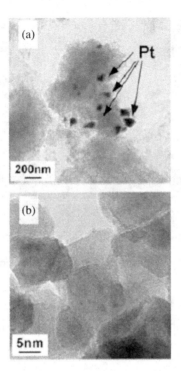

Fig. 6.15. Transmission electron micrographs illustrating the degree of Pt sintering on (a) Al_2O_3 and (b) CZY after treatment at 800°C in air (Nagai *et al.*, 2006).

the step edge, where it reacts with additional oxygen and BaO to form $Ba(NO_3)$. The adsorbed NO_2 is believed to be the critical intermediate during NO_x reduction. Figure 6.16 is an STM image illustrating the Pt{111} surface with BaO_2 on the terraces and NO_2 species at the step edges.

6.6.2.2 *Consiglio Nazionale delle Ricerche (Italy)*

Deganello and coworkers from the CNR/INSM (Palermo) are using sol-gel methods to enhance the thermal stabilities of Pt-Pd/CeO_2-ZrO_2 catalysts and thereby extend the window for lean NO_x reduction. In addition, they are developing Au/Al_2O_3-CeO_2-based NO_x reduction catalysts with enhanced thermal stabilities and higher N_2 selectivities.

Pio Forzatti from the *Politecnico di Milano* (see Site Reports-Europe on the International Assessment of Research in Catalysis by Nanostructured Materials website, www.wtec.org/private/catalysis; password required,

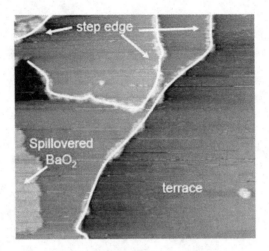

Fig. 6.16. Scanning tunneling microscopy images of the oxidized Ba/Pt{111} surfaces with BaO$_2$ on the terraces.

contact WTEC) described research to better understand mechanisms for NO$_x$ selective catalytic reduction (SCR) with NH$_3$.

6.6.3 *CO$_2$ reduction*

While a number of reactions of CO$_2$ are thermodynamically possible, many are kinetically hindered, therefore, catalytic conversion of CO$_2$ typically requires an energy-intensive, high-temperature process (~1000 K) and the use of hydrogen as a reductant.

$$CO_2 + 3H_2 \leftrightarrow CH_3OH + H_2O, \quad \Delta G° = -9.1 \text{ kJ/mol} \quad (6.9)$$
$$CO_2 + 4H_2 \leftrightarrow CH_4 + 2H_2O, \quad \Delta G° = -130.7 \text{ kJ/mol} \quad (6.10)$$
$$CO_2 + 3H_2 \leftrightarrow \tfrac{1}{2} C_2H_5OH + 3/2\, H_2O, \quad \Delta G° = -48.7 \text{ kJ/mol} \quad (6.11)$$
$$CO_2 + 7/2 H_2 \leftrightarrow \tfrac{1}{2} C_2H_6 + 2\, H_2O, \quad \Delta G° = -96.4 \text{ kJ/mol} \quad (6.12)$$
$$CO_2 + CH_4 \leftrightarrow 2CO + 2H_2, \quad \Delta G° = 170.4 \text{ kJ/mol}. \quad (6.13)$$

The Sabatier reaction (Equation (6.10) is among the most thermodynamically favorable processes; it has been investigated by the US National Aeronautics and Space Administration (NASA) as a means of converting CO$_2$ into hydrocarbon fuels for long interplanetary trips. Transition metals including Ru, Ni, and Fe dispersed on supports like Al_2O_3 catalyze this reaction but have relatively low activity and require temperatures in excess of 400°C. These high temperatures diminish the overall efficiency of the process. The dry reforming of CO_2 with CH_4 (Equation (6.13)) has been extensively investigated in a number of countries, with the goal of developing catalysts that give high yields of syngas without significant deactivation due to coking. A variety of types of supported metal catalysts have been explored, including Ni and noble metal catalysts.

Electrochemical processes offer an attractive strategy for the reduction of CO_2 especially when driven by electricity produced directly or indirectly from solar energy or other carbon-neutral sources (e.g., nuclear energy). Electrochemical potentials for the reduction of CO_2 to C_1 chemicals are listed in Table 6.1. A number of metals and semiconductors are capable of catalyzing these types of reactions. For example, electrodes containing metal phthalocyanines have been reported to reduce CO_2 to formic acid; however, the rates were low (Meshitsuka et al., 1974). Iron, Co, and Ni macrocyclic compounds, including polypyrrole/Schiff-base complexes of Ni, are also capable of reducing CO_2, typically to formate (Dhanasekaran et al., 1999; Ogata et al., 1995; Beley et al., 1986). The reduction of CO_2 to CH_4, C_2H_4 and alcohols has been demonstrated at fairly high yields using metallic copper electrodes and aqueous electrolytes (Hoshi and Hori, 2003); however, the catalytic activities were low and the electrodes quickly deactivated.

6.6.3.1 University of Messina (UNIME; Italy)

Gabriele Centi and coworkers at the University of Messina (see Site Reports-Europe on the International Assessment of Research in Catalysis

Table 6.1. Potentials for electrochemical reduction of CO_2.

Reaction	$E°$ (V_{SHE})
$CO_2(g) + 2H^+(aq) + 2e- \rightarrow HCOOH(aq)$	−0.11
$CO_2(g) + 2H^+(aq) + 2e- \rightarrow CO(g) + H_2O(l)$	−0.10
$CO_2(g) + 4H^+(aq) + 4e- \rightarrow CH_2O(aq) + H_2O(l)$	−0.028
$CO_2(g) + 6H^+(aq) + 6e- \rightarrow CH_3OH(aq) + H_2O(l)$	+0.031
$CO_2(g) + 8H^+(aq) + 8e- \rightarrow CH_4(g) + 2H_2O(l)$	+0.17

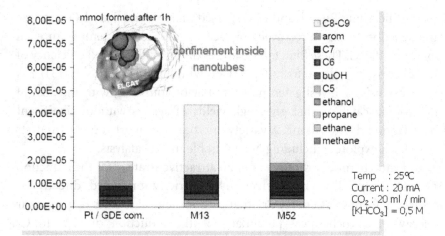

Fig. 6.17. Electrocatalytic conversion of CO_2 into hydrocarbons and alcohols with efficiencies near 30% at ambient conditions have been reported by Centi and coworkers. This figure illustrates production rates for several electrocatalysts including metals confined in nanotubes (courtesy of G. Centi).

by Nanostructured Materials website, www.wtec.org/private/catalysis; password required, contact WTEC) have demonstrated several electrocatalysts including Cu-based materials for the reduction of CO_2 into long chain hydrocarbons and alcohols using an electrochemical cell that resembles a proton exchange membrane fuel cell (Centi et al., 2007; Perathoner et al., 2007). The cells are operated at ambient conditions. As shown in Figure 6.17, a variety of products were formed, including large amounts of aromatic and C8-C9 hydrocarbons. Efficiencies near 30% were reported for this particular process.

6.6.3.2 University of Tsukuba (Japan)

Researchers at the University of Tsukuba (see Site Reports-Asia on the International Assessment of Research in Catalysis by Nanostructured Materials website, www.wtec.org/private/catalysis; password required, contact WTEC) are developing bifunctional oxides for the reduction of CO_2 with alcohols and diols to produce organic carbonates. The bifunctional oxides possess acid and base sites to facilitate insertion of CO_2 into the O-H bond of the alcohol.

6.6.3.3 Institute of Coal Chemistry of the Chinese Academy of Sciences

Researchers at the Institute of Coal Chemistry of the Chinese Academy of Sciences are investigating the use of a variety of reductants including NH_3 to produce a variety of chemicals.

6.7 Summary

Given that global energy needs will continue to be satisfied primarily using fossil resources for the foreseeable future, the importance of catalysis for the production of liquid transportation fuels and hydrogen and for emissions control cannot be overstated. New challenges including those associated with carbon dioxide (CO_2) management also present important opportunities. The WTEC panelists were privileged to witness a great breadth and depth of important energy-related catalysis research in both Europe and Asia. Other observations of the panel are summarized below.

6.7.1 *Project highlights: Energy-centered catalysis R&D*

6.7.1.1 *European Union*

- Modeling of the interactions of MoS_2 and oxide supports for HDS catalysts (Raybaud and coworkers, Institut Français du Pétrole, France)
- Correlations of Co particle size and FTS activity, and use of NO as calcining agent (de Jong and coworkers, University of Utrecht, Debye Institute, The Netherlands)
- Identification of the influence of CeO_2 faceting on WGS activity (Trovarelli and coworkers, University of Udine/CNR, Italy)
- Descriptions of Ba oxide morphologies on Pt NO_x storage materials (Bowker and coworkers, Cardiff University, UK)
- Electrocatalytic reduction of CO_2 to hydrocarbons (Centi and coworkers, University of Messina, Italy).

6.7.1.2 *Asia*

- Continued development of nanostructured Au catalysts (Haruta and coworkers, Tokyo Metropolitan University Department of Applied Chemistry, Japan)

- Use of electrochemical methods (Xu and coworkers, Tsinghua University, China) and plasma methods (Liu and coworkers, Tianjin University, China) to produce nanostructured catalysts
- Development of DCFC as an alternative to coal combustion (Li and coworkers, Tianjin University, China).

6.7.2 *Regional characteristics of catalysis R&D for improving fossil energy production*

6.7.2.1 *European Union*

- Research is highly integrated both nationally and across the European Union
- Academic research is viewed as the driver for innovation and depended upon by industry; thus, there is work on fundamental as well as applied issues
- There are substantial investments in hydrotreating, catalytic cracking, and environmental catalysis.

6.7.2.2 *Asia*

- Research in China is addressing immediate national issues and consequently tends to be highly applied; in some cases it might be better characterized as development
- Applied work in China tends to be focused on utilization of indigenous resources (e.g., coal gasification and FTS), with growing activities in environmental catalysis; much more coal gasification and FTS research than in the United States
- There is emphasis in China on large-scale demonstrations; many of these operations are coordinated with more fundamental work
- Research in Japan and Korea appear to be of similar character and funding as research being carried out in the United States.

References

Aneggi, E., Llorca, J., Boaro, M., and Trovarelli, A. (2005). Surface-structure sensitivity of CO oxidation over polycrystalline ceria powders, *J. Catal.*, 234, p. 88.

Armor, J. N. (1999). The multiple roles for catalysis in the production of H_2, *Appl. Catal. A: Gen.*, 176, pp. 159–176.

Arrouvel, C., Breysse, M., Toulhoat, H., and Raybaud, P. (2004a). Effects of P_{H2O}, P_{H2S}, P_{H2} on the surface properties of anatase–TiO_2 and γ-Al_2O_3: A DFT study, *J. Catal.*, 226, p. 260.

Arrouvel, C., Breysse, M., Toulhoat, H., and Raybaud, P. (2006). A density functional theory comparison of anatase (TiO_2)- and γ-Al_2O_3-supported MoS_2 catalysts, *J. Catal.*, 232, p. 161.

Arrouvel, C., Digne, M., Breysse, M., Toulhoat, H., and Raybaud, P. (2004b). Effects of morphology on surface hydroxyl concentration: A DFT comparison of anatase–TiO_2 and γ-alumina catalytic supports, *J. Catal.*, 222, p. 152.

Bartholomew, C. H. (1990). Recent technological developments in Fischer–Tropsch catalysis, *Catal. Lett.*, 7, pp. 303–316.

Beley, M., Collin, J. P., Ruppert, R., and Sauvage, J. P. (1986). Electrocatalytic reduction of carbon dioxide by nickel cyclam2+ in water: Study of the factors affecting the efficiency and the selectivity of the process, *J. Am. Chem. Soc.*, 108, p. 7461.

Bergwerff, J. A., van de Water, L. G. A., Visser, T., de Peinder, P., Leliveld, B. R. G., de Jong, K. P., and Weckhuysen, B. M. (2005). Spatially resolved Raman and UV-visible-NIR spectroscopy on the preparation of supported catalyst bodies: Controlling the formation of $H_2PMo_{11}CoO_{40}^{5-}$ inside Al_2O_3 pellets during impregnation, *Chem. Eur. J.*, 11, p. 4591.

Bevilacqua, M., Montini, T., Tavagnacco, C., Vicario, G., Fornasiero, P., and Graziani, M. (2006). Influence of synthesis route on morphology and electrical properties of $LaNi_{0.6}Fe_{0.4}O_3$, *Solid State Ionics*, 177, pp. 2957–2965.

Bezemer G. L., Bitter, J. H., Kuipers, H., Oosterbeek, H., Holewijn, J. E., Xu, X., Kapteijn, F., vanDillen, A. J., and de Jong, K. P. (2006). Cobalt particle size effects in the Fischer–Tropsch reaction studied with carbon nanofiber supported catalysts, *J. Am. Chem. Soc.*, 128, p. 3956.

Brenner, J. R., and Thompson, L. T. (1994). Characterization of HDS/HDN active sites in cluster-derived and conventionally-prepared sulfide catalysts, *Catal. Today*, 21, p. 101.

Cabello, C. I., Botto, I. L., and Thomas, H. J. (2000). Anderson type heteropolyoxomolybdates in catalysis: 1. $(NH_4)_3[CoMo_6O_{24}H_6] \cdot 7H_2O/g$-$Al_2O_3$ as alternative of Co-Mo/g-Al_2O_3 hydrotreating catalysts, *Appl. Catal. A*, 197, p. 79.

Cao, L., Chen, J., Liu, S., and Li, Y. (2004). Hydrogen from stepwise reforming of methane: A process analysis, *Stud. Surf. Sci. Catal.*, 147, pp. 103–108.

Capon, A., and Parsons, R. (1973). Oxidation of formic acid at noble metal electrodes. III. Intermediates and mechanism on platinum electrodes, *J. Elec. Chem. Inter. Elec.*, 45(2), pp. 205–231.

Carnell, P. J. H. (1989). *Catalyst Handbook*, ed, Twigg, M. V. (Wolfe Publishing, London), p. 191.

Carrette, L., Friedrich, K. A., and Stimming, U. (2000). Fuel cells: Principles, types, fuels, and applications, *Chem. Phys. Chem.*, 1(4), pp. 162–193.

Centi, G., Perathoner, S., Wine, G., and Gangeri, M. (2007). Electrocatalytic conversion of CO_2 to long carbon-chain hydrocarbons, *Green Chem.*, 9, p. 671.

Chen, J., Li, X., Li, Y., and Qin, Y. (2003). Production of hydrogen and nanocarbon from direct decomposition of undiluted methane on high-nickeled Ni-Cu-alumina catalysts, *Chem. Lett.*, 32, pp. 424–425.

Chen, J., Li, Y., Li, Z., and Zhang, X. (2004). Production of CO_x-free hydrogen and nanocarbon by direct decomposition of undiluted methane on Ni–Cu–alumina catalysts, *Appl. Catal. A: Gen.*, 269, pp. 179–186.

Choplin, A., Leconte, M., Basset, J. M., Shore, S. G., and Hsu, W.-L. (1993). Metallic catalysts starting from heteropolymetallic clusters: Surface organometallic chemistry and Fischer–Tropsch activity of an Fe-Os system, *J. Mol. Catal.*, 21, p. 389.

Chotisuwan, S., Wittayakun, J., Lobo-Lapidus, R. J., and Gates, B. C. (2007). MgO-supported cluster catalysts with Pt-Ru interactions prepared from Pt_3Ru_6 $(CO)_{21}(l_3-H)(l-H)_3$, *Catal. Lett.*, 115, p. 99.

Costa, D., Arrouvel, C., Breysse, M., Toulhoat, H., and Raybaud, P. (2007). Edge-wetting effects of γ-Al_2O_3 and anatase-TiO_2 supports by MoS_2 and CoMoS active phases: A DFT study, *J. Catal.*, 246, p. 325.

Costa, V., Geantet, C., Digne, M., and Marchand, K. B. (2007). Understanding the role of glycol-type additives in the improvement of HDT catalyst performances, *Proc. 234th ACS Nat. Meeting*, Boston.

Costello, C. K., Yang, J. H., Law, H. Y., Wang, Y., Lin, J.-N., Marks, L. D., Kung, M. C., and Kung, H. H. (2003). On the potential role of hydroxyl groups in CO oxidation over Au/Al_2O_3, *Appl. Catal. A: Gen.*, 243, pp. 15–24.

Dhanasekaran, T., Grodkowski, J., Neta, P., Hambright, P., and Fujita, E. (1999). p-rerphenyl-sensitized photoreduction of CO_2 with cobalt and iron porphyrins. Interaction between CO and reduced metalloporphyrins, *J. Phys. Chem. A*, 103, pp. 7742–7748.

Digne, M., Sautet, P., Raybaud, P., Euzen, P., and Toulhoat, H. (2002). Hydroxyl groups on γ-alumina surfaces: A DFT study, *J. Catal.*, 211, p. 1.

Digne, M., Sautet, P., Raybaud, P., Euzen, P., and Toulhoat, H. (2004). Use of DFT to achieve a rational understanding of acid–basic properties of γ-alumina surfaces, *J. Catal.*, 226, p. 54.

Dzwigaj, S., Arrouvel, C., Breysse, M., Geantet, C., Inoue, S., Toulhoat, H., and Raybaud, P. (2005). DFT makes the morphologies of anatase-TiO_2 nanoparticles visible to IR spectroscopy, *J. Catal.*, 236, p. 245.

Energy Information Administration (2007). Emissions of Geenhouse gases in the United States 2006: Flowchart and notes, www.eia.doe.gov/oiaf/1605/ggrpt/flowchart.html.

Energy Information Administration (2008). International energy outlook 2008 report, www.eia.doe.gov/oiaf/ieo/highlights.html.

Gates, B. C. (1995). Supported metal clusters: Synthesis, structure, and catalysis, *Chem. Rev.*, 95, p. 511.

Gates, B. C. (2000). Supported metal cluster catalysts, *J. Mol. Catal.*, 163, p. 55.

Goguet, A., Burch, R., Chen, Y., Hardacre, C., Hu, P., Joyner, R. W., Meunier, F. C., Mun, B. S., Thompsett, D., and Tibiletti, D. (2007). Deactivation mechanism of a $Au/CeZrO_4$ catalyst during a low-temperature water gas shift reaction, *J. Phys. Chem. C*, 111(45), pp. 16927–16933.

Gunardson, H. (1998). *Industrial Gases in Petrochemical Processing* (Marcel Dekker, Inc., New York).

Haruta, M. (2002). Catalysis of gold nanoparticles deposited on metal oxides, *CATTECH*, 6, p. 102.

Haruta, M., and Date, M. (2001). Advances in the catalysis of Au nanoparticles, *Appl. Catal. A: Gen.*, 222, p. 427.

Haruta, M., Kobayashi, T., Sano, H., and Yamada, N. (1987). Novel gold catalysts for the oxidation of carbon monoxide at a temperature far below 0°C, *Chem. Lett.*, 2, p. 405.

Haruta, M., Tsubota, S., Kobayashi, T., Kageyama, H., Genet, M. J., and Delmon, B. (1993). Low-temperature oxidation of carbon monoxide over gold supported on titanium dioxide, α-ferric oxide, and cobalt tetraoxide, *J. Catal.*, 144, p. 175.

He, C., Zhao, N., Du, X., Shi, C., Ding, J., Li, J., and Li, Y. (2006). Low-temperature synthesis of carbon anions by chemical vapor deposition using a nickel catalyst supported on aluminum, *Scripta Materialia*, 54, pp. 689–693.

Hoshi, N., and Hori, Y. (2003). Electrochemical reduction of carbon dioxide on kinked stepped surfaces of platinum inside the stereographic triangle, *Electroanal. Chem.*, 540, p. 105.

Intergovernmental Panel on Climate Change of the United Nations (2007) *Climate Change 2007* (Cambridge University Press, Cambridge), www.ipcc.ch.

Ishida, T., and Haruta, M. (2007). Gold catalysts towards sustainable chemistry, *Angew. Chem. Int. Ed.*, 46, p. 7154.

Karpenko, A., Leppelt, R., Cai, J., Plzak, V., Chuvilin, A., Kaiser, U., and Behm, R. J. (2007). Deactivation of a Au/CeO_2 catalyst during the low-temperature water–gas shift reaction and its reactivation: A combined TEM, XRD, XPS, DRIFTS, and activity study, *J. Catal.*, 250, pp. 139–150.

Kim, C. H., and Thompson, L. T. (2005). Deactivation of Au/CeOx water gas shift catalysts, *J. Catal.*, 230, p. 66.

Kochloefl, K. (1997). *Handbook of Heterogeneous Catalysis*, Eds. Ertl, G., Knözinger, J., and Weitkamp, J. (VCH, Weinheim).

Kolmakov, A., and Goodman, D. W. (2001). Scanning tunneling microscopy of gold clusters on $TiO_2(110)$: CO oxidation at elevated pressures, *Surf. Sci.*, 490, pp. L597–L601.

Li, D., Yang, C., Qi, H., Zhang, H., Li, W., Sun, Y., and Zhong, B. (2004). Higher alcohol synthesis over a La promoted $Ni/K_2CO_3/MoS_2$ catalyst, *Catal. Commun.*, 5, p. 605.

Li, D., Yang, C., Zhao, N., Qi, H., Li, W., Sun, Y., and Zhong, B. (2007). The performances of higher alcohol synthesis over nickel modified K_2CO_3/MoS_2 catalyst, *Fuel Proc. Tech.*, 88, p. 125.

Li, L., Wu, G., and Xu, B. (2006). Electro-catalytic oxidation of CO on Pt catalyst supported on carbon nanotubes pretreated with oxidative acids, *Carbon*, 44, pp. 2973–2983.

Liotta, L. F., DiCarlo, G., Pantaleo, G., Venezia, A. M., Deganello, G., Merlone Borla, E., and Pidria, M. (2007a). Combined CO/CH_4 oxidation tests over Pd/Co_3O_4 monolith catalyst: Effects of high reaction temperature and SO_2 exposure on the deactivation process, *Appl. Catal. B.*, 75, pp. 182–188.

Liotta, L. F., Dicarlo, G., Venezia, A. M., Deganello, G., Merlone Barla, E., and Pidria, M. (2007b). Pd promoted Co_3O_4 catalysts for CH_4 emissions abatement: Study of SO_2 poisoning effect, *Top. Catal.*, 42–43, pp. 425–428.

Marchand, K., Tarret, M., Normand, L., Kasztelan, S., and Cseri, T. (2002). Monitoring of the particle size of MoSx nanoparticles by a new microemulsion-based synthesis, *Stud. Surf. Sci. Catal.*, 143, pp. 239–245.

Martin, C., Lamonier, C., Fournier, M., Mentré, O., Harlé, V., Guillaume, D. and Payen, B. (2005). Evidence and characterization of a new decamolybdocobaltate cobalt salt: An efficient precursor for hydrotreatment catalyst preparation, *Chem. Mater.*, 17, p. 4438.

Meshitsuka, S., Ichikawa, M., and Tamaru, K. (1974). Electrocatalysis by metal phthalocyanines in the reduction of carbon dioxide, *J. Chem. Soc. Chem. Commun.*, pp. 158–159.

Montini, T., Condó, A. M., Hickey, N., Lovey, F., De Rogatis, L., Fornasiero, P., and Graziani, M. (2007a). Embedded Rh (1 wt%)@Al_2O_3: Effects of high temperature and prolonged aging under methane partial oxidation conditions, *App. Catal. B: Environmental*, 73, pp. 84–97.

Montini, T., De Rogatis, L., Gombac, V., Fornasiero, P., and Graziani, M. (2007b). Rh(1%)@Ce$_x$Zr$_{1-x}$O$_2$-Al$_2$O$_3$ nanocomposites: Active and stable catalysts for ethanol steam reforming, *App. Catal. B: Environmental*, 71, pp. 125–134.

Nagai, Y., Hirabayashi, T., Dohmae, K., Takagi, N., Minami, T., Shinjoh, H., and Matsumoto, S. (2006). Sintering inhibition mechanism of platinum supported on ceria-based oxide and Pt-oxide-support interaction, *J. Catal.*, 242(1), pp. 103–109.

National Academy of Engineering of the National Academies (2008). *Grand Challenges for Engineering*, www.engineeringchallenges.org/cms/8996/9221.aspx.

Ogata, T., Yanagida, S., Brunschwig, B. S., and Fujita, E. (1995). Mechanistic and kinetic studies of cobalt macrocycles in a photochemical CO_2 reduction system: Evidence of Co-CO_2 adducts as intermediates, *J. Am. Chem. Soc.*, 117, pp. 6708–6716.

Othmer, K. (1994). *Encyclopedia of Chemical Technology*, vol. 12, 4th Ed. (John Wiley & Sons, Inc., New York).

Parsons, R., and Vandernoot, T. (1988). The oxidation of small organic-molecules. A survey of recent fuel-cell related research, *J. Electroanal. Chem.*, 257(1–2), pp. 9–45.

Pena, M. A., Gomez, J. P., and Fierro, J. L. G. (1996). New catalytic routes for syngas and hydrogen production, *Appl. Catal. A: Gen.*, 144, pp. 7–57.

Perathoner, S., Gangeri, M., Lanzafame, P., and Centim, G. (2007). Nanostructured electrocatalytic Pt-carbon materials for fuel cells and CO_2 conversion, *Kin. Catal.*, 48, p. 877.

Qian, W., Liu, T., Wei, F., Wang, Z., Wang, D., and Li, Y. (2003a). Carbon nanotubes with large cores produced by adding sodium carbonate to the catalyst, *Carbon.*, 41, pp. 2683–2686.

Qian, W., Liu, T., Wei, F., Wang, Z., and Li, Y. (2003b). What causes the carbon nanotubes collapse in a chemical vapor deposition process, *J. Chem. Phys.*, 118(2), pp. 878–882.

Raybaud, P. (2007). Understanding and predicting improved sulfide catalysts: Insights from first principles modeling, *Appl. Catal.*, 322, p. 76.

Sakurai, H., Tsubota, S., and Haruta, M. (1993). Hydrogenation of CO_2 over gold supported on metal oxides, *Appl. Catal. A: Gen.*, 102, p. 125.

Schweiger, H., Raybaud, P., Kresse, G., and Toulhoat, H. (2002). Shape and edge site modifications of MoS_2 catalytic nanoparticles induced by working conditions: A theoretical study, *J. Catal.*, 207, pp. 76–87.

Tibiletti, D., Meunier, F. C., Goguet, A., Reid, D., Burch, R., Boaro, M., Vicario, M., and Trovarelli, A. (2006). An investigation of possible mechanisms for the water–gas shift reaction over a ZrO_2-supported Pt catalyst, *J. Catal.*, 244, p. 183.

Topsøe, H., Clausen, B. S., and Massoth, F. E. (1996). *Hydrotreating Catalysis: Science and Technology* (Springer, Berlin).

Tsubota, S., Haruta, M., Kobayashi, T., Ueda, A., and Nakahara, Y. (1991). Preparation of highly dispersed gold on titanium and magnesium oxide, *Stud. Surf. Sci. Catal.*, 63, p. 695.

US Department of Energy (2008). *Biomass Multi-Year Program Plan* (Office of the Biomass Program, Energy Efficiency and Renewable Energy, Washington), www1.eere.energy.gov/biomass/pdfs/biomass_program_mypp.pdf.

US Environmental Protection Agency (1998). *Toxicological review of hexavalent chromium*, Cas No. 18540-29-9, www.epa.gov/iris/toxreviews/0144-tr.pdf.

Venezia, A. M., Murania, R., Pantaleo, G., and Deganello, G. (2007a). Hydrodesulfurization cobalt-based catalysts modified by gold, *Gold Bull.* 40/2, pp. 130–134.

Venezia, A. M., Murania, R., Pantaleo, G., and Deganello, G. (2007b). Pd and PdAu on mesoporous silica for methane oxidation: Effect of SO_2, *J. Catal.*, 251, pp. 94–102.

Wang, H., Wu, Y., and Xu, B. (2005). Preparation and characterization of nanosized anatase TiO_2 cuboids for photocatalysis, *Appl. Catal. B: Environmental*, 59, pp. 139–146.

Wu, G., Chen, Y., and Xu, B. (2005a). Remarkable support effect of SWNTs in Pt catalyst for methanol electrooxidation, *Electrochem. Commun.*, 7, pp. 1237–1243.

Wu, Y., Liu, H., and Xu, B. (2007a). Solvothermal synthesis of TiO_2: Anatase nanocrystals and rutile nanofibers from $TiCl_4$ in acetone, *App. Organ. Chem.*, 21, pp. 146–149.

Wu, Y., Liu, H. I., Xu, B., Zhang, Z., and Su, D. (2007b). Single-phase titania nanocrystallites and nanofibers from titanium tetrachloride in acetone and other ketones, *Inorg. Chem.*, 46, pp. 5093–5099.

Xin, Z., Wang, H., and Xu, B. Q. (2005a). Remarkable nanosize effect of zirconia in Au/ZrO_2 catalyst for CO oxidation, *J. Phys. Chem. B*, 109, p. 9678.

Xin, Z., Hui, S., and Xu, B. Q. (2005b). Catalysis by gold: Isolated surface Au^{3+} ions are active sites for selective hydrogenation of 1,3-Butadiene over Au/ZrO_2 catalysts, *Angew. Chem. Int. Ed.*, 44, p. 7132.

Yu, Y., and Xu, B. (2006). Shape-controlled synthesis of Pt nanocrystals. An evolution of the tetrahedral shape, *App. Organ. Chem.*, 20, pp. 638–647.

Yu, Y., and Xu, B. (2007). Shape-controlled synthesis of Pt nanocrystals. An evolution of the tetrahedral shape, *App. Organ. Chem.*, 21, p. 209.

Zhang, X., Shi, H., and Xu, B. (2007). Comparative study of Au/ZrO$_2$ catalysts in CO oxidation and 1,3-butadiene hydrogenation, *Catal. Today*, 122, pp. 330–337.

Zhao, D., Wu, G., and Xu, B. Q. (2005). *Chin. Sci. Bull.*, 50, pp. 1846.

Zhao, D., and Xu, B. Q. (2006). Enhancement of Pt utilization in electrocatalysts by using gold nanoparticles, *Angew. Chem. Int. Ed. Eng.*, 45, p. 4955.

7

Applications: Chemicals from Fossil Resources

Vadim V. Guliants

7.1 Introduction

Heterogeneous catalysts enable many chemical transformations of fossil resources into useful products (Thomas and Thomas, 1997; Gates, 1992). Catalysts are responsible for the production of over 60% of all chemicals and are used in some 90% of all chemical processes worldwide (Council for Chemical Research, 1998; Bartholomew and Farrauto, 2005). According to a 2001 discussion of the impact of catalysis on the US economy, "one-third of material gross national product in the US involves a catalytic process somewhere in the production chain" (Morbidelli et al., 2001). Catalyst manufacturing alone accounts for over US$10 billion in sales worldwide in four major sectors: refining, chemicals, polymerization, and exhaust emission catalysts. However, the value derived from catalyst sales is greatly eclipsed by the total value of the products that are produced, i.e., chemical intermediates, polymers, pesticides, pharmaceuticals, and fuels. The global annual impact of catalysis is estimated to be US$10 trillion (Council for Chemical Research, 1998). As we look to the future, heterogeneous catalysis increasingly holds the key to "green chemistry" and the promise of eliminating or at least dramatically curbing pollution from chemical and refining processes (Bowker, 1998). Heterogeneous catalysis is key to many proposed green chemical processes targeted to cut back harmful emissions (Anastas and Warner, 1998). The goal of catalysis for green chemistry is to tailor atomically the structure of the active and selective site in order to convert reactants directly to products without generating by-products that typically end up as waste (van Santen and Neurock, 2006).

The majority of industrial catalysts contain an active component in the form of nanoparticles < 20 nm in size that are dispersed onto high-surface-area supports. The importance of nanoparticles and nanostructure to the performance of catalysts has stimulated wide efforts to develop methods for their synthesis and characterization, making this area of study an integral part of nanoscience (Bell, 2003). This chapter provides recent examples from the literature and from the institutions visited by the WTEC panel of how catalysis performance (i.e., activity and selectivity) is affected by the local size, shape, structure, and composition of catalyst particles employed in four major classes of organic reactions that underlie conversion of fossil resources to chemicals. These four classes are alkylation, dehydrogenation, hydrogenation, and selective oxidation. The chapter also highlights future trends in catalysis by nanostructured materials that are expected to result in detailed understanding of the effects of particle composition, size, and structure on catalyst performance, which will support better green chemistry for the future. Ultimately, the goal of studies in catalysis by nanostructured materials is to develop an understanding of these complex systems to a point where it will be possible to successfully exploit nanoscale phenomena by design to create new heterogeneous catalysts and green catalytic processes.

7.2 Alkylation

The alkylation of aromatic compounds is widely used in the large-scale synthesis of petrochemicals, fine chemicals, and intermediates (Ertl et al., 1997). This reaction consists of the replacement of a hydrogen atom of an aromatic compound by an alkyl group derived from an alkylating agent. If the hydrogen being replaced is on the aromatic ring, the reaction is electrophilic substitution, which requires an acid catalyst. If the hydrogen being replaced is on the side chain of an aromatic molecule, then base catalysts or radical conditions are needed.

Acid catalysts used for alkylation of aromatic hydrocarbons are Brønsted acids containing acidic protons, e.g., acidic halides such as $AlCl_3$ and BF_3, acidic oxides, zeolites, protonic acids — especially sulfuric acid, hydrofluoric acid, phosphoric acid — and organic cation exchange resins. The acidic halides and protonic acids are being rapidly replaced for large-scale applications by solid alkylation catalysts, especially zeolites, because these are much more desirable for environmental reasons. They are non-corrosive and offer additional advantages for controlling selectivity via their shape-selective properties. This brief overview focuses on recent developments related to the synthesis and applications of novel solid catalysts in the alkylation of aromatic compounds in which the nanoscale pore environment of a solid catalyst is exploited to achieve superior activity and shape selectivity.

The periodic mesoporous organosilicas (PMOs) represent one of the recent breakthroughs in the field of materials chemistry of mesoporous nanostructured materials (e.g., Asefa et al., 1999). PMOs are synthesized from the bridged organosilane precursors, $(R'O)_3SiR(OR')_3$ and possess some unique properties that cannot be obtained by other approaches, such as uniformly distributed organic groups in the mesoporous framework with maximum loading of 100%. Unlike fully inorganic variants of ordered mesoporous frameworks, PMOs exhibit a periodic arrangement of the hydrophobic organic layers (e.g., ethylene and benzene) and hydrophilic silica layers within the pore walls (see Figure 7.1) (Yang et al., 2004; Xia et al., 2005; Yang et al., 2005). Incorporation of heteroatoms in these PMOs provides a new approach for the synthesis of bifunctional mesoporous materials with crystal-like pore wall structure and novel catalytic properties in epoxidation (Ti), ammoxidation (Ru), and alkylation (Al) reactions (e.g., Wahab and Ha, 2005, and references therein).

In studies where tetrahedral Al species were incorporated into the PMO frameworks during synthesis of ethylene- (Yang et al., 2004; Xia et al., 2005) and benzene-bridged PMOs (Yang et al., 2005), the resultant PMO phases with both mesoscale (2.3–3.0 nm pore diameters, 4.6–5.1 nm d_{100}-spacings, and 2.1–2.6 nm thick walls) and molecular-scale periodicity

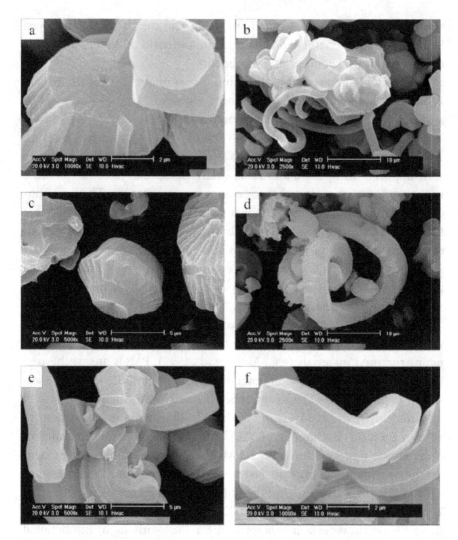

Fig. 7.1. Representative SEM images of ethylene-containing hybrid mesoporous organosilica (Figure 3 from Xia *et al.*, 2005).

possessed significantly improved hydrothermal stability, due to increased hydrophobic character and the incorporation of tetrahedral aluminum in the mesoporous framework. It was found that the ethylene-bridged mesoporous framework generated a greater amount of acid sites than the phenylene-bridged network with a similar Si/Al ratio and aluminum coordination environment. Consequently, the Al-containing ethylene-bridged PMO exhibited higher catalytic activity in alkylation of 2,4-di-*tert*-butylphenol with cinnamyl alcohol under identical reaction conditions (Yang et al., 2005). Moreover, periodic arrangement of organic layers in these materials offers a possibility to introduce several types of functional groups and even control their relative spacing on the mesopore surface, which together with metal substitution into their silica framework component may offer a new strategy for controlling multistep catalytic reactions in a single vessel.

Microporous zeolites have been widely used in industry as solid-acid catalysts for a variety of alkylation reactions. Significant mass transport limitations to and from the active sites located in micropores severely limit their performance (Davis, 2002). To overcome these limitations, various strategies have been successfully pursued, such as the synthesis of nanosized zeolites (early work on colloidal zeolites Y and A by Schoeman et al., 1994), ultralarge-pore zeolites and zeolite analogs (VPI-5, JDF-20, UTD-1, CIT-5, SSZ-53, ECR-34, ITQ-21, etc.; see Corma, 2003) and ordered mesoporous materials (MCM-41, SBA-15, FSM-16, etc.; see Wan and Zhao (2007)). However, the use of these materials is rather limited, due to the difficulty of separating nanosized zeolite crystals from the reaction mixture, the complexity of the templates used for the synthesis of ultra large-pore zeolites, and the relatively low thermal and hydrothermal stability of ordered mesoporous materials. More recently, mesoporous zeolites from nanosized carbon templates have also been reported (e.g., Jacobsen et al., 2000), but their industrial applications are still limited by the complexity of the synthetic procedure involved and the hydrophobicity of the carbon templates.

Recently, hierarchical mesoporous zeolite beta (Beta-H) templated from a mixture of small organic ammonium salts and mesoscale cationic polymers has been reported that possesses dual porosity due to the presence of zeolite nanocrystals forming a mesoporous structure, which provides significantly improved mass transport behavior in alkylation catalysis (Xiao et al., 2006). This route involves a one-step hydrothermal synthesis, and the templated mixture is homogeneously dispersed in the synthetic gel. Importantly, these novel hierarchical zeolites exhibit excellent catalytic properties as compared to conventional zeolite beta. Beta zeolite is generally synthesized from a small organic template of tetraethylammonium hydroxide (TEAOH). Hierarchical mesoporous Beta zeolite (Beta-H) is crystallized in the presence of TEAOH and a mesoscale cationic polymer, polydiallyldimethylammonium chloride (PDADMAC).

Scanning and transmission electron microscopy (SEM and TEM) images of Beta-H (Figure 7.2) reveal the presence of zeolite particles about 600 nm in size and hierarchical mesoporosity in the 5–40 nm range. Partial connections may be observed between these hierarchical pores; these are beneficial for the mass transfer of reactants and products in catalysis. Beta-H showed a high activity and selectivity as an alkylation catalyst

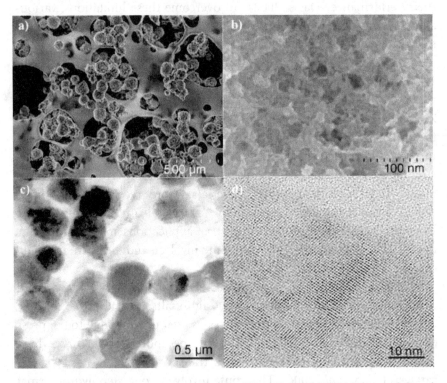

Fig. 7.2. Electron microscopy images of calcined Beta-H: (a)–(b) SEM images at low and high magnification, respectively; the separation between each marker represents 5 μm and 100 nm, respectively); (c)–(d) TEM images at low and high magnification, respectively (Figure 2 from Xiao et al., 2006).

(Figure 7.3), as well as a long catalyst life relative to the sample of conventional Beta zeolite. The similarities of Beta-H to conventional Beta zeolite in terms of Si/Al ratios, aluminum distribution, and acidic strength, as well as the larger particle size of Beta-H than that of Beta zeolite indicate that the higher catalytic activity of Beta-H in the model alkylation reaction is related to the hierarchical mesoporosity in Beta-H, which is important for improving the mass transport of the reactants and products in the alkylation of benzene with propan-2-ol.

The presence of hierarchical mesoporosity in the Beta-H sample is attributed to the use of the molecular and aggregated cationic polymer PDADMAC. The molecular weight of the cationic polymer lies in the range 10^5–10^6, and its size is estimated at 5–40 nm, which is in good agreement with the dimensions of the mesopores obtained from high-resolution TEM studies (Figure 7.2(d)). The cationic polymers could effectively interact with negatively charged inorganic silica species in alkaline media, resulting in the hierarchical mesoporosity. The addition of a greater

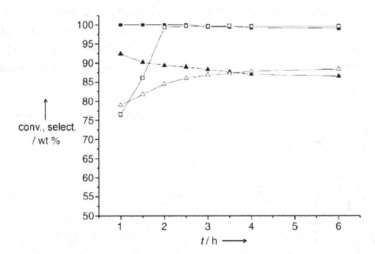

Fig. 7.3. Catalytic conversion (conv. [wt%]) and selectivity (select. [wt%]) in the alkylation of benzene with propan-2-ol as a function of reaction time (reaction conditions: 200°C; 4:1 benzene/propan-2-ol; pressure: 2.0 MPa, WHSV: 10 h^{-1}). Conversion on Beta-H (■); selectivity on Beta-H (□); conversion on Beta zeolite ( ); selectivity on Beta zeolite (△) (Figure 3 from Xiao et al., 2006).

amount of cationic polymer in the synthetic gel yields Beta zeolite with larger mesoporosity, indicating the controllable mesoporosity of the zeolite sample. The synthesis of hierarchical mesoporous zeolites is not limited to the Beta variety, which was obtained through use of TEAOH with PDADMAC, but other mixtures of organic amine salts and cationic polymer templates may be used if they effectively interact with inorganic species in alkaline media under conditions to crystallize the zeolites. For example, hierarchical mesoporous ZSM-5 zeolite (ZSM-5-H) was obtained using a mixture of tetrapropylamine hydroxide and dimethyl-diallyl ammonium chloride acrylamide copolymer (10 wt%). Particularly, the use of the hierarchical mesoporous ZSM-5-H in the catalytic cracking of 1,3,5-triisopropylbenzene showed that it is much more active a catalyst than conventional ZSM-5 under the same reaction conditions (Xiao et al., 2006).

The WTEC team visited the group of Avelino Corma at the Institute of Chemical Technology (Instituto de Tecnología Química or ITQ) in Valencia, Spain. This group has reported novel ultra-large-pore silicogermanate zeolite ITQ-33, which exhibits straight large pore channels with circular openings of 18 rings along the c-axis interconnected by a bidirectional system of 10-ring channels, yielding a structure with very large micropore volume (Figure 7.4, left; Corma et al., 2006; see also ITQ site report in Site Reports-Europe on the International Assessment of Research

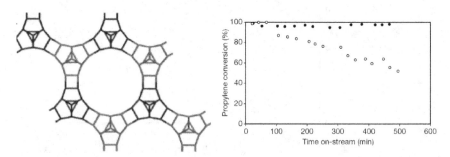

Fig. 7.4. (*Left*) view of the ITQ-33 structure along [001], showing the 18-ring structure (oxygen atoms have been removed for clarity); (*right*) percentage propylene conversion as a function of time on-stream for ITQ-33 and Beta; ITQ-33 is represented by filled circles; Beta is represented by open circles. T = 1250 C; P = 3.5 MPa; WHSV = 12 h-1; the benzene/propylene molar ratio is 3.5. (Corma et al., 2006, Figures 3 [part] and 4, respectively).

in Catalysis by Nanostructured Materials website, www.wtec.org/private/catalysis; password required, contact WTEC). Although the synthesis conditions of ITQ-33 are easily accessible, they are not typical for silicogermanate zeolite analogs and were identified using high-throughput synthesis techniques. ITQ-33 possesses the surface acidity capable of catalyzing with very high-activity, interesting catalytic reactions, such as alkylation of benzene with propylene to produce the industrially relevant cumene, while giving an extremely low yield of the undesirable n-propylbenzene (below 0.01% yield at 99% conversion). When working at very low contact time (weight hourly space velocity or WHSV = 12 h^{-1}), ITQ-33 decays much more slowly than the Beta zeolite commercially used at present (Figure 7.4, right). Activity is maintained after at least five reaction regeneration cycles (540°C in air). ITQ-33 is very active for alkylation and transalkylation of alkyl aromatics, as well as for dealkylation of bulky alkylaromatics containing one or two condensed aromatic rings.[1]

ITQ-33 is of interest for producing more diesel and less gasoline while maintaining the propylene and butene yield, during catalytic cracking of vacuum gasoil. This issue is of importance, given that diesel has higher mileage efficiency than gasoline and given the growing imperative to save fuel and restrict CO_2 emissions. This predicted catalytic behavior would be a consequence of the very large pores that should increase the diesel yield, and the existence of the 10-ring connecting pores that allow diffusion and cracking of gasoline molecules, producing C_3 and C_4 olefins. ITQ-33 gives a cracking conversion higher than Beta and close to that of a USY (ultrastable Y) zeolite. Furthermore, ITQ-33 produces more diesel, less gasoline and propylene/propane, and isobutene/isobutene ratios much higher than does USY, and very similar to results for Beta zeolite. Used together, ITQ-33 and ZSM-5 have an excellent cooperative effect, giving a much higher diesel and propylene yield, with lower gasoline yield, than does the combination of USY and ZSM-5 zeolites used today. Zeolite ITQ-33 may thus transform the field of zeolite catalysis, if its stability and economics can be further improved.

---

[1] In summarizing work performed abroad, the author has used some of the groups' own descriptions of their work.

The Corma group has demonstrated that it is possible to synthesize amorphous microporous molecular sieves with different pore dimensions and topologies predefined by the size and shape of the organic structure-directing agent (Corma and Díaz-Cabañaz, 2006). From the energetic diagram for the different steps occurring during the synthesis of zeolites (Figure 7.5), it appears possible to synthesize in an analogous way stable amorphous microporous molecular sieves with pore dimensions predefined by the size and shape of the organic structure directing agent. These synthesis steps are (a) an induction period, (b) nucleation, in which viable nuclei are formed, and (c) growth of the nuclei to form zeolite crystals. Following the reaction steps given in Figure 7.5, the Corma group synthesized and isolated, before nucleation occurred, stable amorphous microporous materials. These new microporous stable materials can be considered as amorphous zeolite precursors (ZP) containing zeolitic nuclei that are too small to show crystallinity by different spectroscopic techniques. These novel nanosized zeolitic materials have several advantages over conventional zeolites, i.e., they can always be obtained in high yields, with shorter synthesis times, and in larger compositional ranges.

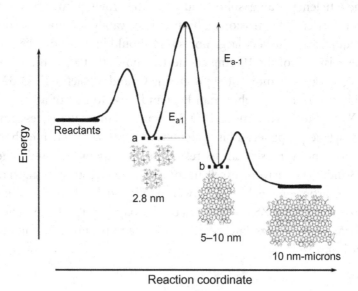

Fig. 7.5. Energetics of zeolite crystallization (Figure 1 from Corma and Díaz-Cabañaz, 2006).

Corma and Díaz-Cabañaz also reported ZSM-12 (monodimensional 12-ring channel), NU-87 (bidimensional 10-ring pore), and ITQ-21 (three-dimensional 12-ring channels) zeolite precursors that possessed micropore diameters and pore volumes similar to corresponding well-crystallized zeolites. These zeolite precursors correspond to partially arranged amorphous versions of the zeolite with already a large amount of Si–O–Si bonds formed, but with still an important fraction of internal defects, as was evidenced by ^{29}Si MAS NMR. ^{27}Al MAS NMR showed that Al in calcined zeolite precursors is in tetrahedral coordination and accounts for the Brønsted acidity observed by pyridine-adsorption/desorption measurements. These acid sites are active and selective for catalyzing alkylation and cracking reactions of hydrocarbons. The catalytic results show that the microporous amorphous molecular sieve ZPITQ-21 gives the same catalytic activity as the corresponding large pore zeolite (12-R) ITQ-21 for cracking 1,3-diisopropylbenzene (DIPB), which can easily diffuse into 12-ring pore zeolites. However, for cracking the bulkier 1,3,5-triisopropylbenzene (TIPB), ZPITQ-21 is more active than ITQ-21, showing the presence of slightly larger pores, probably due to incompletely formed cages and cavities in the former material. The above observation is consistent with the distribution of the trimethylpentane isomers (TMP) obtained during alkylation of 2-butene with isobutane. ZPITQ-21 is more selective towards the bulkier 2,2,4-TMP and 2,2,3-TMP (which have the highest octane numbers) than the corresponding ITQ-21. It has to be remarked that the molar ratio R (R = (2,2,4-TMP + 2,2,3-TMP)/(2,3,4-TMP + 2,3,3-TMP)) has been associated with shape selectivity effects in zeolites, this ratio being higher for zeolites with larger pores.

7.3 Dehydrogenation and Hydrogenation

There are a number of significant developments related to synthesis and application of novel metal nanoparticle and other nanostructured catalysts in two broad classes of heterogeneously catalyzed organic reactions, dehydrogenation and hydrogenation. A wide spectrum of both dehydrogenation and hydrogenation reactions is catalyzed by noble metal catalysts (e.g., palladium [Pd] and platinum [Pt]) dispersed on metal oxide supports, whereas dehydrogenation reactions, including *oxidative* dehydrogenation reactions (ODH), are also catalyzed by supported transition metal oxides, such as vanadium pentoxide (V_2O_5), molybdenum trioxide (MoO_3), and chromic acid (Cr_2O_3).

7.3.1 Dehydrogenation

Recently, the Somorjai group at the University of California, Berkeley, demonstrated that the rate of cyclohexene hydrogenation and dehydrogenation is influenced by the symmetry of the platinum single crystal faces (McCrea and Somorjai, 2000). They concluded that the maximum turnover rate of hydrogenation appeared at lower temperature than the dehydrogenation, and the maximum hydrogenation rate was higher on Pt (111) while lower on Pt (100) faces compared to the rate of dehydrogenation. The phenomenon was traced back to the difference of reaction mechanisms on the two different surfaces and serves as an excellent test reaction occurring with different rates over different crystal faces.

The same group reported for the first time the preparation of catalysts with well-defined shaped metal (platinum, gold, silver, etc.) nanoparticles in the pore system of ordered mesoporous silicas. In this method a colloid solution of metal nanoparticles protected by organic molecules is used as template for the synthesis of SBA-15 (Kónya et al., 2002) and/or MCM-41 (Kónya et al., 2003) mesoporous silicates. A novel type of catalyst composed of well-shaped (cubic) metal particles and mesoporous matrix (pore size 7–8 nm) was prepared and characterized by various physicochemical techniques. The platinum nanoparticles have well-defined shapes, such as cubic, tetrahedral, cubo-octahedral, etc. The cubic particles have (100) faces, the tetrahedral particles display (111) faces, and the cubo-octahedral particles have both type of faces. In the most favored case, the platinum nanoparticles might be embedded in the pores of mesoporous silica structures, such as MCM-41 and SBA-15. The catalysts prepared in this way may mimic the catalytic processes studied in two dimensions in three dimensions.

Such nanostructured catalytic materials confined to cavities and pores of regular nanoscale dimensions are the subject of continuing interest because they have unique size-dependent catalytic properties, which are significantly different than those of the corresponding bulk catalysts

(Rupprechter, 2007). The group of researchers at the State Key Laboratory of Catalysis, Dalian Institute of Chemical Physics, China, which the WTEC panel visited, has demonstrated that Pt nanoparticle catalysts for methylcycloxane (MCH) dehydrogenation may be prepared by a "one-pot" encapsulation into the mesopore channels of ordered mesoporous SBA-15 silica (Chen et al., 2007). The fabrication of metallic Pt nanoparticles with controllable size and shape has become an important topic in nanotechnology owing to their unique catalytic performance. The use of ordered mesoporous materials as hosts to limit the growth of nanostructured materials in their pores is a highly promising approach to stabilize metal nanoparticles against undesirable aggregation and sintering. Preparation of Pt nanoparticles or nanowires in porous materials typically involves at least two or three steps to restrict Pt nanoparticles in the mesoporous matrix. Chen et al. (2007) developed a novel one-pot approach to directly introduce Pt nanoparticles into the mesochannels of SBA-15, which required no preexisting mesoporous host and Pt nanoparticles, and no extra reduction process of the platinum precursor (Figure 7.6).

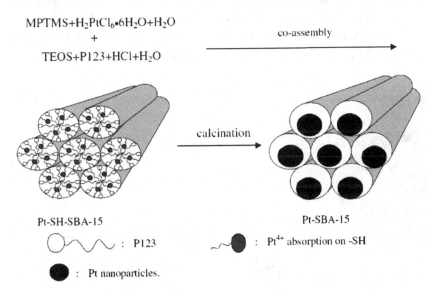

Fig. 7.6. Preparation of highly dispersed Pt nanoparticles in the SBA-15 mesochannels (Scheme 1 from Chen et al., 2007).

This co-assembly method is based on the I$^+$M$^-$S$^+$ scheme for the synthesis of mesoporous host (Zhao et al., 1998). The positively charged surfactants such as protonated block copolymers [S$^+$] and cationic inorganic oxide precursors [I$^+$] are assembled together through the mediator [M$^-$]. 3-mercaptopropyltrimethoxysilane [MPTMS] with thiol groups are added to modify the cationic precursors [I$^+$] in order to confine the Pt nanoparticles inside the mesochannels because platinum ions are easy to combine with the thiol groups through strong chemical bonds. The mediator in this case could be the anionic platinum complex and chloride ions.

Figure 7.7 shows the results from the total synthesis route to obtain monodisperse Pt nanoparticles in SBA-15. The direct co-assembly of MPTMS and tetraethylorthosilicate (TEOS) is preferred over the commonly used post synthesis grafting between surface silanols and functional silylation agents, which provides for a more homogeneous distribution of organic ligands in the framework. After adsorption of platinum ions onto the thiol groups, platinum sulfide analogues can be decomposed into metallic platinum and sulfur oxides when heating in air providing a homogeneous distribution of metallic Pt nanoparticles in the mesochannels of SBA-15 after calcination at 550°C to remove the template (Figure 7.6).

These nanocomposites exhibited higher catalytic activity and stability than conventional Pt/SiO$_2$ catalysts in the methylcyclohexane

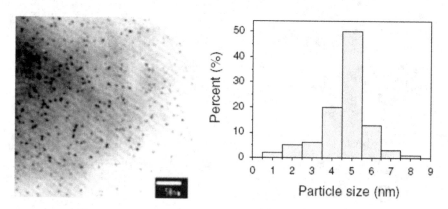

Fig. 7.7. TEM images and particle size distribution of Pt–SBA-15 (Figure 5 from Chen et al., 2007).

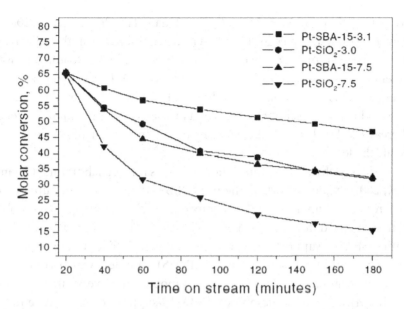

Fig. 7.8. Methylcyclohexane conversion by Pt–SBA-15 and conventional Pt–SiO$_2$ catalysts at atmospheric pressure, 300°C and WHSV of 27.1 as a function of time on-stream (Figure 6 from Chen et al., 2007).

dehydrogenation (Figure 7.8). All samples showed nearly 100% selectivity to toluene over the entire experimental run. The initial MCH conversion was ca. 65.0% for all catalysts, which then decreased with time on-stream. Pt confined in SBA-15 displayed higher dehydrogenation conversion and stability than that supported on SiO$_2$ under the same reaction conditions. The smaller particle size and more homogeneous dispersion of Pt in SBA-15 may result in its higher catalytic stability in MCH dehydrogenation as the confinement of ordered mesochannels restricts further growth of Pt nanoparticles during the dehydrogenation reaction.

In an ongoing collaboration, the groups of Can Li at the State Key Laboratory of Catalysis of the Dalian Institute of Chemical Physics, China, and Rutger van Santen at Schuit Institute of Catalysis at the Technical University of Eindhoven (see site reports in Site Reports-Asia and Site Reports-Europe on the International Assessment of Research in Catalysis by Nanostructured Materials website, www.wtec.org/private/catalysis; password required, contact WTEC), investigated Fe substitution into the inorganic walls of SBA-15 (Li et al., 2008; Guan et al., 2007; Li et al., 2006). They determined that Fe location had a strong influence on the selectivity to dehydrogenation and dehydration of ethanol. At low Fe loading, Fe was present as isolated species in the amorphous SBA-15 silica and formed aggregated clusters of iron oxide at high Fe loading. The isolated Fe species possessed Brønsted acidity that resulted in selective formation of ethylene, whereas the Fe clusters were efficient in the formation of ethylene and acetaldehyde.

Besides dehydrogenation catalysis, the same collaborators (Zhang et al., 2007) explored further the use of SBA-15 as a catalytic support to prepare a novel interfacial hybrid epoxidation catalyst by a new immobilization method for homogeneous catalysts. Their approach involved coating SBA-15 support with an organic polymer film containing active sites. The titanium silsesquioxane (TiPOSS) complex, which contains a single-site titanium active center, was immobilized successfully by *in situ* copolymerization on a mesoporous SBA-15-supported polystyrene polymer. The resulting hybrid materials exhibit attractive textural properties, such as highly ordered mesostructure, large specific surface area (> 380 m^2g^{-1}) and pore volume (≥ 0.46 cm^3g^{-1}), and high activity in the epoxidation of alkenes. In the epoxidation of cyclooctene with *tert*-Bu hydrogen peroxide (TBHP), the hybrid catalysts have rate constants comparable to that of their homogeneous counterparts and can be recycled at least seven times. These immobilized catalysts can also catalyze the epoxidation of cyclooctene with aqueous H_2O_2 as the oxidant. In two-phase reaction media, the catalysts show much higher activity than their homogeneous counterparts due to the hydrophobic environment around the active centers. They behave as interfacial catalysts due to their multifunctionality, i.e., the hydrophobicity of polystyrene and the polyhedral oligomeric silsesquioxanes (POSS), and the hydrophilicity of the silica and the mesoporous structure. The simultaneous immobilization of homogeneous catalysts on two conventional supports (inorganic solid and organic polymer) has been shown to provide novel heterogeneous catalytic ensembles with attractive textural properties, tunable surface properties, and optimized environments around the active sites.

Several novel synthesis approaches to surface immobilization of homogeneous catalysts in porous or high surface area supports (Figure 7.9)

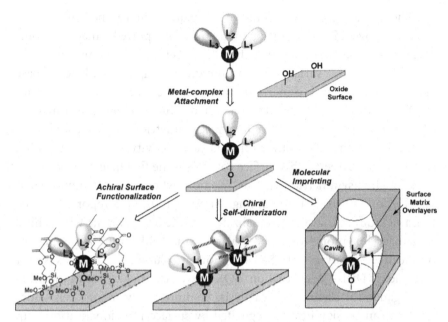

Fig. 7.9. Methods for advanced surface design with supported metal complexes (Figure 1 from Tada and Iwasawa (2006)).

have been reported by the Iwasawa group at the University of Tokyo (Tada and Iwasawa, 2006; see also Site Reports-Asia on the International Assessment of Research in Catalysis by Nanostructured Materials website, www.wtec.org/private/catalysis; password required, contact WTEC). Interfacial chemical bonding of Pd monomers with Pd–P (P: P(O–iPr)$_3$, PMe$_2$Ph, and dppf) and Pd–N (N: tmeda, methylpiperidine, and cyclohexylamine) to SiO$_2$, Al$_2$O$_3$, and TiO$_2$ surfaces was employed to obtain a series of supported Pd complex catalysts for the hydroamination of 3-amino-propanol vinyl ether. The order of the hydroamination activities observed for these catalysts correlated with the pKa values of the SiO$_2$, Al$_2$O$_3$, and TiO$_2$ surfaces. The most ionic bond, Pd–OSi, was favorable for the hydroamination of alkenes, while the Pd–OAl bond with relatively more covalent character did not efficiently promote the reaction. These results show that the chemical bonding with surfaces has a major impact on catalytic activity, and the nature of the support is an important parameter that can be explored to produce new catalytic behavior.

The Iwasawa group, which the WTEC panel visited at the University of Tokyo, discovered chiral self-dimerization of supported vanadium complexes on a SiO_2 surface, which is a novel phenomenon for metal complexes on oxide surfaces. Two V-monomer complexes with Schiff-base ligands spontaneously dimerized via a selective reaction with a surface Si–OH group, and the formed V dimer had a unique chiral conformation, which is highly enantioselective for the asymmetric oxidative coupling of 2-naphthol with 96% conversion, 100% selectivity to 1,1'-binaphthol (BINOL), and 90 ee%. This surface complex is the first heterogeneous catalyst for the asymmetric coupling reaction, whereas the V monomer is inactive for the oxidative coupling. Another technique reported by the Iwasawa group, surface functionalization of the SiO_2 support with achiral 3-methacryloxypropyl-trimethoxysilane, remarkably amplified the enantioselective catalysis of SiO_2-supported Cu–BOX complexes for asymmetric Diels–Alder reaction. BOX (bis(oxazoline)) is one of the practical ligands for asymmetric catalysis. Enantioselectivity for Diels–Alder reaction can be significantly regulated by surface functionalization with achiral silane-coupling reagents on SiO_2-supported Cu–BOX complexes.

Molecular imprinting methods have recently showed significant promise in creating template-shaped cavities with memory of template molecules that are reminiscent of artificial enzymes possessing recognition ability for particular substrate molecules (Katz and Davis, 2000). Acid–base catalysts and metal complexes synthesized by molecular imprinting techniques provide promising molecular recognition catalysis with 100% selectivity for a variety of catalytic reactions where natural enzymes cannot be employed (Tada and Iwasawa, 2006).

Molecular imprinting typically consists of several steps: (1) attachment of a metal complex on robust supports, (2) surrounding of the metal complex by a polymer matrix, and (3) production of a shape selective cavity on the metal site in the matrix. Most of the imprinted metal-complex catalysts have been prepared by imprinting in bulk polymers, which discourages the access of reactant molecules to the active sites in the bulk. Moreover, polymers tend to be unstable in organic solvents or under demanding catalytic conditions, such as in the presence of oxidants, at high temperatures, etc. To overcome these limitations, the Iwasawa group designed molecular-imprinted catalysts for oxide-supported metal complexes to produce shape-selective reaction sites by using a ligand on a metal center as a template. A ligand of a supported metal complex not only influences its catalytic activity but also provides an unsaturated, reactive metal site with a ligand-shaped space after removal of a ligand.

The Iwasawa group chose a ligand of the attached metal complex with a similar shape to a reaction intermediate (half-hydrogenated alkyl) of alkene hydrogenation as a template. This strategy to design active and selective catalysts was based on the following five factors for regulation: (1) conformation of ligands coordinated to a rhodium (Rh) atom, (2) orientation of vacant site on Rh, (3) cavity with complementary molecular shape for the reaction space produced after template removal, (4) architecture of the cavity wall, and (5) micropore in inorganic polymer-matrix overlayers stabilizing the active species at the surface. A $P(OCH_3)_3$ ligand was used as a template with a similar shape to one of the half-hydrogenated species of 3-ethyl-2-pentene, which can produce the template (reaction intermediate)-shaped cavity after extraction of the ligand.

The Iwasawa group succeeded in preparing Rh monomer and a pair of Rh monomers by using appropriate precursors on the surfaces. On both surfaces, $Si(OCH_3)_4$ was deposited by chemical vapor deposition and converted into SiO_2-matrix overlayers surrounding the attached Rh complexes via a hydrolysis–polymerization step. Finally, the template ligand, $P(OCH_3)_3$, was extracted from the attached Rh complex in the SiO_2-matrix overlayers yielding the molecularly imprinted Rh-monomer and Rh-dimer catalysts. The homogeneous complexes, $Rh_2Cl_2(CO)_4$ and $RhCl(P(OCH_3)_3)_3$, and the supported species, $Rh_2Cl_2(CO)_4/SiO_2$, showed no activity for alkene hydrogenation at 348 K. On the other hand, the molecular-imprinted catalysts exhibited significant catalytic hydrogenation activities under similar reaction conditions. For example, hydrogenation of 2-pentene on the molecularly imprinted Rh-dimer catalyst was promoted 51 times as compared to that on the supported catalyst. The metal–metal bonding and coordinative unsaturation of the Rh dimer are key factors for the remarkable activity of the imprinted Rh-dimer catalyst. The selectivity for the alkene hydrogenation on the molecularly imprinted catalysts depended on the alkene size and shape which should come into the reaction site in a template cavity in addition to the electronic and geometric effects of the ligands.

The location of the Rh center for alkenes coordination, the conformation of the remaining $P(OCH_3)_3$ ligand, the orientation of the template vacant site on Rh, the template-shaped cavity, the architecture of the cavity wall, and the micropore surrounding the Rh dimer in the SiO_2-matrix overlayers provided active imprinted catalysts for the size- and shape-selective alkene hydrogenation. Therefore, the arrangement of active sites on surfaces by chiral self-dimerization, surface functionalization with achiral reagents, and molecular imprinting provide new powerful approaches for the design of selective catalyst surfaces in three dimensions beyond conventional homogeneous and heterogeneous catalyst systems.

7.3.2 Hydrogenation

The use of well-defined metal nanoparticles (1–10 nm) for catalytic processes is a rapidly growing area (see recent reviews by Astruc et al., 2005; Chandler and Gilbertson, 2006; Tsuji and Fujihara, 2007; Andres et al., 2007). Similar to molecular complexes, metal nanoparticles are efficient and selective catalysts for hydrogenation of olefins and C–C couplings, but also for reactions that are not catalyzed or are poorly catalyzed by molecular species, such as hydrogenation of arenes. However, despite impressive progress in asymmetric catalysis, few colloidal systems have been found to display an interesting activity in this field. Those systems that show promise include Pt(Pd)/cinchonidine for the hydrogenation of ethyl pyruvate and Pd-catalyzed kinetic resolution of racemic substrates in allylic alkylation.

Metal nanoparticle catalysts can be obtained by a variety of methods according to the organic or aqueous nature of the media and the stabilizers used (e.g., Andres et al., 2007). Poly(vinyl)pyrrolidone, PVP, has been the homopolymer most used as stabilizing agent for metallic nanoparticles. Others like cellulose and polysaccharide, polyvinylalcohol, polystyrene, polyacid or poly(vinyl)formamide derivatives, and copolymers and various dendrimers have been also applied for similar purposes. Side-chain functionalized polymers for their applications as stabilizers of metallic nanoparticles are particularly attractive, because the functional groups can interact with the metallic surface (Favier et al., 2007). Favier and colleagues functionalized poly(methyl vinyl ether-co-maleic anhydride) and employed it to stabilize ~3–20 nm sized Pd, Pt, and Rh nanoparticles, which were investigated in catalytic hydrogenation of ethyl pyruvate (Rh nanoparticles) and C–C coupling using phenylboronic and 2-methylnaphthyl-1-yl boronic acids (Pd nanoparticles). The Rh nanoparticle catalysts employed in the hydrogenation reaction were significantly more active than the conventional heterogeneous Rh catalyst and resistant to agglomeration and sintering.

However, organic polymers are not very stable at high temperatures, and the catalytic reactions in solution may be accompanied by some swelling, which can cause significant mass-transfer resistances. Therefore, the use of inorganic supports or porous hosts appears to be more suitable. Pârvulescu et al. (2006) reported SiO_2-embedded Pd, Au, and highly alloyed Pd–Au colloids prepared by sol–gel embedding of presynthesized colloids and their performances in the hydrogenation of several substrates: cinnamaldehyde, 3-hexyn-1-ol, and styrene. These data showed evidence that alloying Pd with Au in bimetallic colloids leads to enhanced activity and most importantly to improved selectivity. Moreover, the combination of the two metals resulted in catalysts that were very stable against poisoning, as was observed for the styrene hydrogenation in the presence of thiophene.

An interesting example of metal nanoparticle systems prepared by chemical vapor deposition of volatile metal organic precursors into the large pores of metal organic frameworks (MOF) has been reported by Hermes et al. (2005). Hermes and coworkers synthesized several metal@MOF systems in which Cu and Pd nanoparticles were grown inside the pores of MOF-5 framework and were found to be active catalysts for methanol production from syngas (Cu@MOF-5) and cyclooctene hydrogenation (Pd@MOF-5), whereas in the case of the Au@MOF-5 system, Au atoms migrated from the pore cavities of MOF-5 to the external surface and aggregated into large 20 nm Au particles, which were inactive in CO oxidation.

Crooks and colleagues at the University of Texas at Austin and Texas A&M University recently investigated a particle size effect for hydrogenation over unsupported Pd nanoparticle catalysts in a size range (1.3–1.9 nm) that has not been widely studied and demonstrated that the rate of hydrogenation of allyl alcohol is a function of the diameter of the Pd nanoparticles (Wilson et al., 2006). Furthermore, kinetic data indicated that this effect is probably electronic in nature for particles having diameters < 1.5 nm, but for larger particles it depends primarily on their geometric properties. The Crooks researchers employed dendrimer-encapsulated nanoparticles (DENs) prepared using dendrimer templates that exert a high degree of control over the size, composition, and structure of catalytically active nanoparticles in the < 3 nm size range. Specifically, sixth-generation, hydroxyl terminated polyamidoamine dendrimers (G6-OH) were used to synthesize Pd DENs containing an average of 55, 100, 147, 200, or 250 Pd atoms (G6-OH [Pd_n]), where n is the average number of atoms per particle).

Alkene hydrogenation occurs via the Horiuchi–Polanyi mechanism, which involves dissociative adsorption of H_2 onto the catalyst surface, followed by stepwise hydrogenation of the C = C double bond (Horiuchi and Polanyi, 1934). Figure 7.10 shows that only the total number of face atoms

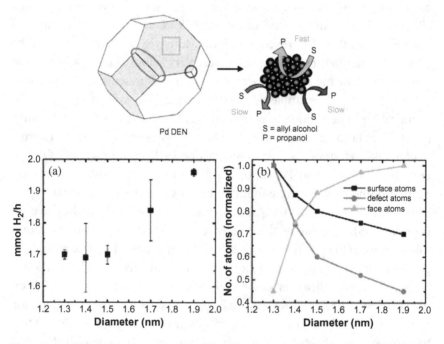

Fig. 7.10. (a) Plot of the rate of hydrogen consumption as a function of particle diameter, (b) plot of the numbers of surface, defect, and face atoms for each particle size. The data are normalized to the largest number of each type of atom (Figure 1 from Wilson et al., 2006).

increases with particle size, while the numbers of surface and defect atoms both decrease. Differences in reaction rates as a function of catalyst size arise from either electronic or geometric effects. For example, as the size of a nanoparticle decreases, its electronic properties change from those of a metal to an insulator and then to something akin to those of a molecule, which modulate the catalytic properties of nanoparticles. Geometric effects are most evident when a reaction requires a specific type of surface atom, because the ratio of defect (vertex and edge) to face atoms changes dramatically as a function of size for < 5 nm diameter particles. While geometric effects have been observed for homogeneous colloidal Pd catalysts for both the Heck and Suzuki coupling reactions, literature reports pertaining to size effects in hydrogenation reactions focus almost exclusively on supported (heterogeneous) Pd catalysts (Wilson et al., 2006).

There is a major conceptual difficulty in that, in the size range where the *electronic* structure of metal particles is changing, there are also major changes in the *geometric* arrangement of atoms on the surface (Bond, 2005; Bond et al., 2006; Bond, 2007). Therefore, it is impossible to assign the catalytic effects of particles of size less than ~5 nm unambiguously to either a geometric or an electronic effect. However, if the size effect persists for metal particles larger than 5 nm, it is more likely due to a requirement of a specific active site, since the changes in electronic structure become insignificant for such larger particles.

In addition to catalyst size, the preparation and structure of nanoparticles are also important factors that must be taken into account when comparing catalytic activity. This is because preparation methods, stabilizing ligands, and polydispersity can lead to activity changes that may mask true particle size effects. For example, Pd nanoparticles having similar diameters, but stabilized by either poly(vinylpyrrolidone) or 1,10-phenanthroline, exhibit very different catalytic activities for the hydrogenation of 1,3-cyclooctadiene (Toshima et al., 2001). The hydrogenation reaction is sensitive to both the electronic and geometric properties of the catalytic Pd nanoparticles, both of which change quickly in the 1.3–1.9-nm diameter size range. The hydrogenation kinetics of allyl alcohol (Wilson et al., 2006) are dominated by electronic effects for the smallest particles (< 1.5 nm diameter) and by geometric effects for larger particles (1.5–1.9 nm diameter). Results of the type described here were enabled by the high degree of monodispersity resulting from the dendrimer templating approach to nanoparticle synthesis.

The synthesis of particles with narrow size distribution and homogeneous physicochemical properties in order to establish reactivity-morphology relationships is a highly desirable but challenging effort. Controlling the shape at the nanoscale translates into ability to control the relative exposure of different crystal facets and the number of atoms on corners and edges and, hence, ability to tune the activity and selectivity of a catalytic system. The application of shape-controlled synthesis of metallic particles in catalysis is a relatively new direction. However, the number of studies using such catalytic systems has been rapidly growing in recent years. Fukuoka *et al.* (2001) reported a direct comparison between catalytic properties of spherical and Pt nanowires synthesized inside a mesoporous template. In the hydrogenolysis of butane, Pt nanowires exhibited a 36 times higher turnover frequency than spherical particles. Moreover, Pt nanowires were found to produce ethane by secondary–secondary carbon bond cleavage, whereas spherical Pt particles were able only to cleave to terminal C–C bonds and then to produce methane and propane. This change of catalytic properties was ascribed to the preferential exposure of (110) planes and/or to more electron-deficient sites. Park *et al.* (2007) reported tetrahedral Rh nanoparticles preferentially exposing the catalytically active (111) planes that were prepared by thermal decomposition of Rh complexes in oleylamine and supported on charcoal. These supported (4.9 ± 0.4) nm Rh particles showed a 5.8- and 109-fold increase in activity for the hydrogenation of anthracene compared to spherical Rh nanoparticles and commercial Rh/C catalyst, respectively.

The catalysis group at the Catholic University of Leuven in Belgium recently reported Au^0 nanocolloids stabilized by polyvinylpyrrolidone (PVP), which displayed high chemoselectivity in the hydrogenation of α, β-unsaturated aldehydes and ketones to allylic alcohols (Mertens *et al.*, 2007). Analogous Pt^0 and Ru^0 nano-colloids were more active than Au^0, but substantially less chemoselective for allylic alcohols. Experimental control over the Au^0 cluster generation provided the opportunity to investigate the size-dependent catalytic behavior of nano-Au^0 and determine the optimum gold cluster size leading to the highest allylic alcohol yields. The optimum cluster size of the Au^0 colloids obtained using

HAuCl$_4$·3H$_2$O as the metal precursor and various PVP/Au ratios was determined for the highest crotyl alcohol yield. At the optimum, the PVP/Au ratio was 6, and based on TEM, nearly spherical Au^0 colloids have a mean diameter of 7 nm.

Telkar et al., (2004) compared the catalytic properties of spherical and cubic Pd particles for hydrogenation of butyne-1,4-diol and of styrene oxide. Very high turnover frequencies were found in both cases using cubic nanoparticles. Moreover, cubic particles were found to be more selective than spherical particles for the hydrogenation of butyne-1,4-diol into but-2-ene-1,4-diol. Similarly, the selectivity in 2-phenyl ethanol for the hydrogenation of styrene oxide was nearly 100% on cubic particles versus less than 50% for conventional catalytic systems. The most detailed study was undertaken by Narayanan and El-Sayed (2004a; 2004b). They used a model reaction, the electron transfer reaction between hexacyanoferrate(III) and thiosulfate ions to study the catalytic properties of cubic, tetrahedral and spherical Pt particles. The particles' reactivity follows the increasing order: cubes < spheres < tetrahedra. The higher reactivity of tetrahedral nanoparticles was related to a higher proportion of corner and edge sites on these particles exposing only {111} faces. Therefore, the morphological control during the synthesis of nanoparticles seems to lead to highly selective catalysts, due to preferential exposure of crystallographic planes. Most of these studies were performed using polymers (polyvinylpyrrolidone, N-isopropylacrylamide, polyacrylate). These organic ligands can modify catalytic properties by affecting the electronic structure of Pt nanoparticles as well as by limiting the accessibility of active sites to reactants and intermediates (Telkar et al., 2004; Sakamoto et al., 2004).

A team of researchers from three French institutions — Institut de Recherches sur la Catalyse et l'Environnement de Lyon, Université Pierre et Marie Curie Paris VI, and Institut Français du Pétrole in Lyon (see Site Reports-Europe on the International Assessment of Research in Catalysis by Nanostructured Materials website, www.wtec.org/private/catalysis; password required, contact WTEC) — successfully employed structure-directing agents and a so-called "seeding-mediated" approach to prepare stable and surfactant-free anisotropic Pd nanoparticles (Figure 7.11). These were investigated in selective hydrogenation of buta-1,3-diene

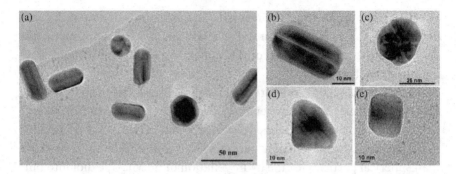

Fig. 7.11. TEM images of anisotropic Pd nanoparticles: (a) nanorods and polyhedra, (b) a typical nanorod with five-fold symmetry, (c) a polyhedron formed from six tetrahedral subunits, (d) a tetrahedral particle with rounded edges, (e) nanocube with rounded edges (Figure 1 from Berhault et al., 2007).

(Berhault et al., 2007). In this method, anisotropic growth of nanoparticles is controlled by the selective adsorption of structure-directing agents, e.g., ions, surfactants, or polymers. The selective adsorption of a structure-directing agent inhibits the growth of particles along a given crystallographic direction through a preferential stabilization of specific faces. The polyol process was extensively used to synthesize Group VIIIb noble metal-based nanostructured catalysts. The polyol process uses a polymer as structure-directing agent (generally, PVP), and reduction is performed by the alcoholic solvent. In aqueous media, surfactants (mainly cetyltrimethylammonium bromide, CTAB) play the role of structure-directing agent. Synthesis in aqueous solutions is particularly attractive in the way that (1) surfactant can be eliminated easily from the surface of the nanoparticles and (2) using a "seeding-mediated" approach, the respective rates of nucleation and growth could be better controlled through the injection of seeds serving as unique centers of nucleation-growth at the beginning of the growth process.

Berhault and coworkers applied this methodology to the synthesis of anisotropic Pd nanoparticles that they employed in the selective hydrogenation of buta-1,3-diene. Selective hydrogenation of unsaturated compounds is a technologically important process in heterogeneous catalysis. Dienes are unwanted by-products of thermal cracking of petroleum cuts. Depending on the target objective, buta-1,3-diene should be converted selectively into but-1-ene in the course of the production of polymers, or into but-2-ene for producing petrochemicals (higher olefins, alcohol), or into gasoline of high octane number after alkylation. Palladium-based catalysts are frequently used for the selective hydrogenation of buta-1,3-diene. This reaction is structure-sensitive, and studies on model catalysts have shown that Pd (110) exhibits a higher selectivity for the formation of butenes than Pd (111) (Silvestre-Alberto et al., 2005; Katano et al., 2003). Similarly, Guo and Madix (1995) have also observed that Pd (100) surfaces are particularly selective for the hydrogenation of dienes into alkenes without further hydrogenation into alkanes.

Berhault and coworkers successfully applied a seeding-mediated approach to synthesize well-defined morphological Pd nanoparticles (nanocubes, nanotetrahedra, nanopolyhedra, and nanorods) using cetyltrimethylammonium bromide (CTAB) as both capping and structure-directing agent (Figure 7.12). Pd nanorods were found to expose preferentially (100) lateral planes due to the selective inhibition by CTAB

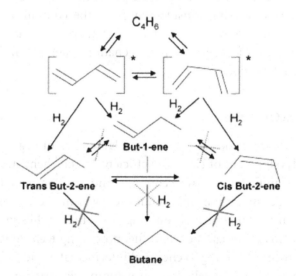

Fig. 7.12. Proposed reaction network for buta-1,3-diene hydrogenation on the nanostructured Pd catalysts. The dotted crosses indicate unfavorable routes for the nanostructured Pd catalysts as compared to the isotropic catalyst, whereas solid crosses indicate prohibited routes (Scheme 1 from Berhault et al., 2007).

of the crystalline growth of {100} facets. After deposition onto α-Al_2O_3, these nanostructured Pd particles were tested in the selective hydrogenation of buta-1,3-diene at 298 K under 20 bars of H_2. As compared to an isotropic Pd catalyst, anisotropic Pd catalysts were found to be highly selective for the hydrogenation of buta-1,3-diene into butenes without further hydrogenation into butane. Moreover, further hydrogenation of but-1-ene into butane over these nanostructured catalysts was reduced, whereas but-2-enes were hardly converted. This would be related to the higher exposition of {100} facets on the Pd nanorod catalyst in the absence of any selective poisoning effect due to CTAB. Finally, an unusually high level of *cis* but-2-ene isomer was observed, suggesting a modification of the mechanism usually accepted for Pd catalysts when a well-defined nanomorphology is achieved. The mechanism for the hydrogenation of buta-1,3-diene on the nanostructured catalysts differs from the one generally assumed on "more classical" isotropic catalysts with a higher *cis/trans* but-2-ene ratio, probably resulting from an interconversion between *syn* and *anti* di-adsorbed butadiene species (Figure 7.12). Therefore, nanostructured Pd catalysts present interesting and new catalytic properties differing considerably from the conventional isotropic catalysts. They are also useful model catalysts in order to bring new mechanistic information for selective hydrogenation reactions under relevant experimental conditions.

7.4 Selective Oxidation

Selective oxidation processes represent a large class of organic reactions where the development of clean and efficient "green chemistry" processes can have a significant positive economic and environmental impact to mitigate the accumulation of greenhouse gases in the atmosphere and other environmental concerns resulting from the world's growing consumption of fossil fuels and current fuel processing techniques. Selective oxidation processes leading to chemical intermediates and fine chemicals have up to now largely relied on stoichiometric reactions employing chromate, permanganate, and renate species. Simple catalytic reactions making use of molecular oxygen or hydrogen peroxide that require minimal energy and produce minimal by-product waste are highly desirable in order to replace stoichiometric reactions that are expensive and environmentally unfriendly.

7.4.1 Selective oxidation catalysis by nanosized gold and other noble metals

A key emerging technology that shows great promise in selective oxidation is catalysis by gold and gold-containing nanoparticles and nanoclusters (Ishida and Haruta, 2007; Hashmi and Hutchings, 2006). Gold was earlier considered to be chemically inert and regarded as a poor catalyst. However, when gold is highly dispersed on metal oxides as small nanoparticles with diameters of less than 10 nm (and preferably in the 1.5–3 nm range), it becomes a highly active oxidation catalyst for many reactions, such as CO oxidation and propylene epoxidation in the gas phase (Haruta, 2003). Most recently, polymer-supported gold has also shown great promise as a catalyst for green liquid-phase selective oxidation processes (Miyamura et al., 2007). They achieved higher catalytic reactivity for the polymer-supported catalysts than for gold supported on metal oxides via simultaneous optimization of the support, solvents, and size of gold particles.

Over the last decade, following the pioneering work by Prati and Rossi (1998), much attention also has been paid to the development of liquid-phase oxidation processes employing gold nanoparticle catalysts. The ultimate objective is to carry out green liquid-phase oxidations at atmospheric pressure and room temperature, either in aqueous media or under solvent-free conditions, and using air as oxidant. Gold nanoparticles supported on activated carbon or metal oxides are active for liquid-phase selective oxidation of alcohols and polyols into the corresponding aldehydes, ketones, or carboxylic acid with molecular oxygen in aqueous media (Corma and Serna, 2006; Ketchie et al., 2007). In most cases, gold catalysts showed high catalytic activity with much higher selectivity at lower temperature and stability than palladium and platinum catalysts. The selective oxidation of glucose by gold supported on activated carbon or metal oxides has been a very active research area, as the transformation of readily available glucose to valuable gluconic acid is of great importance (e.g., Comotti et al., 2005). In fact, gold catalysts supported on activated carbon and Al_2O_3 were efficient under moderate conditions (e.g., atmospheric pressure, using molecular oxygen as oxidant, and reaction temperatures of 303–333 K). In the case of Au on activated carbon, turnover frequencies (TOF) reached high values (12,500 to 25,000 h^{-1}), and high selectivities to gluconic acid were observed ranging from 92 to 99% by varying the kind of activated carbon.

Au nanoparticles (2–5 nm) supported on nanocrystalline CeO_2 (ca. 5 nm) gave a particularly active, selective, and recyclable catalyst for the oxidation of alcohols using molecular oxygen at atmospheric pressure under solvent-free and base-free conditions (e.g., Abad et al., 2006). The TOF value for oxidation of 1-phenylethanol (12,480 h^{-1}) exceeded that of Pd nanoparticles supported on hydroxyapatite (9,800 h^{-1}) (Mori et al., 2004). The Au/CeO_2 showed a much higher selectivity for the oxidation of allylic alcohols to unsaturated ketones, e.g., yielding 1-octen-3-one as a main product in the presence of Au/CeO_2 with a selectivity of 93% in the oxidation of 1-octen-3-ol. The Pd/CeO_2 displayed a lower selectivity of 58% because it promoted isomerization and hydrogenation of the C=C bond to yield saturated ketones as a by-product (Abad et al., 2006). Moreover, Au supported on activated carbon and Au/Al_2O_3 preferentially oxidized the hydroxy group in 2-aminopropanol to yield 2-aminopropanoic acid and avoided the oxidation of the amino group (Biella et al., 2002), indicating an attractive chemoselectivity in liquid-phase aerobic oxidations.

Recently, the interest of the catalysis community has shifted from activated carbon and metal oxides to organic polymers as a stabilizing agent or a solid support for gold nanoparticles in liquid-phase oxidations. Tsunoyama et al. (2005) reported that small gold clusters 1.3 nm in diameter stabilized by hydrophilic poly(N-vinylpyrrolidone) (Au:PVP) catalyzed the oxidation of benzyl alcohol into benzaldehyde as the main product and benzoic acid as a by-product in water at 303 K using molecular oxygen as oxidant. They indicated that the TOF value was inversely proportional to the diameter of the Au nanoparticles in the 3–6 nm range and, further, increased significantly for Au nanoparticles smaller than 3 nm. The disadvantage of the Au:PVP system is that the Au catalyst was essentially present as a homogeneous catalyst, which complicates its separation from the reaction medium.

Gel-type polyacrylic resin, poly-2-(methylthio)ethyl methacrylate-N,N-dimethyl acrylamide-N,N-methylene bisacrylamide (MTEMA-DMAA-4–8), was successfully employed to prepare recyclable supported catalysts containing well-defined gold nanoparticles (Burato et al., 2005). The thioether (R-S-R) group was used as a coordination site for a Au precursor prior to its reduction to Au^0 with $NaBH_4$. The gel-type resin had small pores with a mean diameter of 2.5 nm, while the size of the Au nanoparticles was 2.2 nm, suggesting that the cavities in the polymer gel prevented the Au nanoparticles from aggregating. The gold nanoparticles supported on the gel-type polymer catalyzed the oxidation of pentanol to pentanal (25% conversion) and of pentanal to pentanoic acid (95% conversion). These performances, however, were still inferior to that of Au/CeO_2. Miyamura et al. (2007) developed recyclable polystyrene-based copolymer-microencapsulated gold nanocatalysts (PI-Au), in which gold nanoparticles were stabilized by benzene rings in the polystyrene moiety. These PI-Au catalysts were active at room temperature when using the appropriate combination of water, organic solvent, and base. Recently, the Haruta group (see Tokyo Metropolitan University site report, Site Reports-Asia on the International Assessment of Research in Catalysis by Nanostructured Materials website, www.wtec.org/private/catalysis; password required, contact WTEC) has developed a simple, patented, technique to prepare polymer catalysts by using surface functional groups on an anion-exchange resin as a reducing agent (Ishida et al., 2008). The obtained Au/polymer catalysts exhibit a higher TOF (27,000 h^{-1}) than the Au/activated carbon catalyst the group of Rossi employed in the oxidation of glucose to gluconic acid in water (Comotti et al., 2005).

The Hutchings group at Cardiff University in the UK (see Site Reports-Europe on the International Assessment of Research in Catalysis by Nanostructured Materials website, www.wtec.org/private/catalysis; password required, contact WTEC) had previously studied solvent effects on the selective oxidation of cyclohexene over Au on activated carbon, where the highest selectivity for cyclohexene oxide (50%) was obtained in 1,2,3,5-tetramethylbenzene, while 2-cyclohexen-1-one was obtained as a main product (78% selectivity) in 1,3,5-trimethylbenzene at 353 K (Hughes et al., 2005). The Kobayashi group at the University of Tokyo has shown that further optimization of reaction media could lead to the use of milder conditions, such as room temperature (Miyamura et al., 2007). They also examined the solvent-free oxidation of 1-phenylethanol at 433 K under almost the same reaction conditions (at atmospheric pressure) as employed by the Corma group (Abad et al., 2006). The TOF value for PI-Au (20,000 h^{-1}) exceeded that for the Au/CeO$_2$ catalyst (12,480 h^{-1}) reported by Abad et al., 2006.

Functionalized organic polymers are highly promising not only as a support to stabilize small gold nanoparticles but also as promoters of higher catalytic activity as compared to conventional gold catalysts supported on metal oxides and activated carbon. Despite the recent progress, the reaction mechanism involving the functionalized group in the polymer backbone and the size effect of gold nanoparticles both remain unclear. In the case of Au/CeO$_2$, the proposed reaction mechanism involves cationic gold species and Ce^{3+}, and the oxygen-deficient sites of CeO$_2$ are assumed to play an important role for physisorption of molecular oxygen. Additionally, the production of small gold clusters with diameters of less than 2 nm, which is a critical size for a dramatic change in electronic structure, is still an emerging and exciting area of research (Bond et al., 2006). These small clusters composed of less than 300 atoms can be tuned more pronouncedly by choosing support materials ranging from metal oxides and activated carbon to polymers. The success of the Kobayashi group in controlling the size of gold particles down to 1 nm through entrapment in a polymer particle represents an important step toward the goal of sustainable chemistry.

For Au nanoparticles supported on metal oxides, a general synthetic strategy developed by the Stucky group at the University of California, Santa Barbara, to prepare and disperse Au nanoparticles even on acidic and hydrophobic supports appears to be highly promising (Zheng and Stucky, 2006). In this method the narrow size distribution of Au nanoparticles is generated by employing weaker reducing agents than the commonly used NaBH$_4$, followed by capping the stable monodisperse Au nanoparticles by long-chain alkyl thiols to yield hydrophobic nanoparticles. A homogeneous loading of capped Au nanoparticles is created by utilizing their weak interactions with oxide surfaces in aprotic solvents, which is then locked in place by calcination to remove the surface capping groups. Zheng and Stucky reported 3.5–8.2 nm Au particles by selecting different solvents and controlling the reaction temperatures. However, it remains to be seen if this method could be adapted to prepare even smaller Au nanoparticles (< 3 nm in diameter).

Lastly, great success has been achieved in a variety of catalytic processes by combining two metallic elements in bimetallic catalysts, such as the platinum-iridium (Pt-Ir) system for petroleum reforming, platinum-tin (Pt-Sn) for alkane dehydrogenation, and the nickel-gold (Ni-Au) system for steam-reforming of alkanes (Bond, 2005). The recent interest in gold as a catalytic element has led to a series of studies on gold-containing bimetallic combinations (Bond et al., 2006). In particular, the palladium-gold (Pd-Au) system has been found more effective than either component by itself in a number of selective oxidations, including those of reducing sugars, alkenes, and hydrogen (to form hydrogen peroxide rather than water). In the macroscopic state, palladium and gold form a continuous range of solid solutions, and it was expected that chemical methods for preparing nanoscale bimetallic particles would also lead to microalloys. Surprisingly, detailed structural examination has shown that this is not always the case; instead, a core-shell structure often occurs, in which a gold core is surrounded by a palladium shell (Edwards et al., 2005). The beneficial catalytic effect is therefore obtained by the Au core exerting some kind of modifying influence on the surface Pd atoms, which this has not yet received a theoretical explanation.

Intriguing results for bimetallic AuPd catalysts supported on activated carbon in the aqueous-phase oxidation of glycerol have been reported by the Davis group at the University of Virginia (Ketchie et al., 2007). They prepared bimetallic catalysts by depositing Au from solution onto commercial 3 wt.% Pd/C catalyst. According to EXAFS data, the AuPd catalyst at low Au loading contained Au "skin" on the Pd core, whereas at high Au loading, gold deposited as highly dispersed monometallic clusters on the support. Another bimetallic AuPd was prepared by the Au sol method and gave a catalyst in which Au almost completely covered Pd. However, no significant rate enhancement was found for the bimetallic catalysts compared with monometallic Au after the rates of glycerol oxidation were properly normalized to the surface metal concentration. Pd does appear to have an important influence on the selectivity of the glycerol oxidation reaction. Kinetic studies revealed that Pd in either a bimetallic particle or a physical mixture with Au decreased the H_2O_2 formation in the reaction. Since H_2O_2 is correlated with the formation of C–C cleavage products, the presence of Pd with Au increased the selectivity of the glycerol oxidation to the formation of glyceric acid. These results (Ketchie *et al.*, 2007; Bond *et al.*, 2006; Edwards *et al.*, 2005) further suggested that bimetallic and multimetallic Au-based catalysts represent another highly promising and relatively unexplored dimension of Au catalysis in selective oxidation reactions.

7.4.2 *Selective oxidation of lower alkanes by bulk mixed metal oxides*

Selective oxidation catalysis is of vital importance to society, providing about a quarter of all organic chemicals and intermediates used to make consumer goods and the industrial products that rely on them (Grasselli, 2002). The goal to use less-expensive feedstock materials has spawned the development of new and improved heterogeneous catalysts for C2-C4 conversions. In the selective oxidation and ammoxidation of C3 feedstocks, significant efforts are underway to use propane feeds in processes that produce acrylic acid and acrylonitrile over processes that have traditionally relied on propene feeds catalyzed by promoted (multicomponent) bismuth molybdates for the production of acrolein and acrylonitrile. The production of acrylic acid (currently from acrolein) and acrylonitrile has reached a scale that approaches the equivalent of nearly 1 kg for every human being on Earth each year (Pyrz *et al.*, 2008).

Among various candidate systems investigated for selective (amm)oxidation of propane, the so-called "M1/M2" phase system with the bulk Mo-V-Nb-Te-O composition appears to be the most viable for commercial implementation. This system contains two crystalline phases (M1 and M2); the synergy between them has been claimed to be critical for the achievement of optimal selectivity during propane conversion (Holmberg *et al.*, 2004; Baca *et al.*, 2005). M1 is the majority phase in this catalyst and has an orthorhombic polyhedral network-type molybdenum bronze structure with a framework similar to that of Mo_5O_{14} and related bronzes (Figure 7.13). This phase can be described with the generic formula $\{TeO\}_{1-x} \cdot (Mo,V,Nb)_{10}O_{28}$, where the $\{TeO\}$ component is intercalated into framework channels.

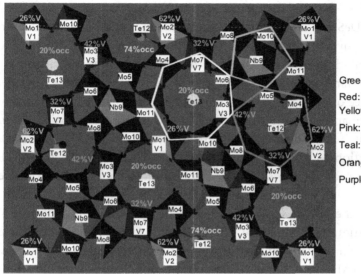

Fig. 7.13. Bulk *ab* planes of M1 phase (DeSanto *et al.*, 2004). The refined orthogonal structure had the space group *Pba2*, a formula unit $Mo_{7.8}V_{1.2}NbTe_{0.937}O_{28.9}$, and lattice parameters a = 21.134(2) Å, b = 26.658(2) Å, and c = 4.0146(3) Å with Z = 4 (Figure 1 from Pyrz *et al.*, 2008).

The M2 phase is optimally present in amounts of roughly 20–30% and is an orthorhombically distorted hexagonal tungsten bronze (HTB)-type structure (DeSanto *et al.*, 2004). In analogy to the generic description of the M1 formula, the formula for M2 can be written as $\{TeO\}_{2-x} \cdot (Mo,V,Nb)_6O_{18}$. This M1/M2 composite is both active and selective, providing up to 62 mol.% acrylonitrile yield in propane ammoxidation. These conversions require a trifunctional active site where the activation of the highly stable paraffin C–H bond on a V(V) site is followed by the abstraction of α-H and a subsequent sacrificial insertion of oxygen (oxidation possibly involving a O = Mo = O species) or nitrogen (ammoxidation requiring co-fed NH_3 that was proposed to adsorb and form O = Mo = NH species) on the catalyst surface (Grasselli, 2005). The structural and chemical complexity of the M1/M2 system may provide a site-isolated complex trifunctional active site for selective conversion (Grasselli, 2005). The insertion step requires that lattice oxygen be indirectly replenished, and the overall process involves a complex interplay of electrons and protons at the active surface.

Although the crystal structures of the M1 and M2 phases have been reported (DeSanto et al., 2004), the detailed description of molybdenum (Mo), vanadium (V), and niobium (Nb) locations and their roles in the overall catalytic behavior of the Mo-V-Te-Nb-O system is still rather limited. For instance, although only the M1 phase is able to activate propane, the presence of M2 was shown to improve the selectivity of the M1 phase during propane oxidation to acrylic acid and propane ammoxidation to acrylonitrile (e.g., Holmberg et al., 2004; Baca et al., 2005). The propane ammoxidation to acrylonitrile is proposed to occur via the propylene intermediate, where the M1 phase is responsible for propane oxidative dehydrogenation to propylene and its further ammoxidation to acrylonitrile, while the M2 phase was proposed to be more efficient in propylene ammoxidation to acrylonitrile. However, the cooperation between these two phases in propane ammoxidation is still poorly understood because the M1 and M2 phases employed in these previous studies were often prepared by different synthesis techniques (e.g., slurry evaporation versus solid-state) and possessed different chemical compositions.

7.4.3 Catalytic behavior of Mo-V-(Te-Nb)-O M1 phase catalysts

The Ueda group (Catalysis Research Center at the University of Hokkaido; see Site Reports-Asia on the International Assessment of Research in Catalysis by Nanostructured Materials website, www.wtec.org/private/catalysis; password required, contact WTEC) and the Guliants group at the University of Cincinnati recently reported compositionally simpler Mo-V-O, Mo-V-Te-O, and more complex Mo-V-Te-Nb-O M1 phase catalysts prepared by hydrothermal synthesis (Watanabe and Ueda, 2006; Shiju and Guliants, 2007). Despite close similarities in the bulk crystal structure and chemical composition, the Mo-V-(Te-Nb)-O catalysts displayed significant differences in catalytic performance in propane ammoxidation to acrylonitrile (Korovchenko et al., 2008). The M2 phases were inactive in propane ammoxidation, whereas the Mo-V-O M1 phase displayed the highest activity among all M1 phases studied but modest selectivity to acrylonitrile that was similar to that of the Mo-V-Te-O M1 phase catalyst. However, the presence of Nb together with Te in the M1 phase resulted in a dramatic drop in the activity in propane and propylene ammoxidation, accompanied by significant enhancement in the selectivity to acrylonitrile.

7.4.4 On cooperation of M1 and M2 phases in propane ammoxidation

Several groups have reported improvement in the acrylonitrile yield as a result of the cooperation of the M1 and M2 phases in the Mo-V-Te-Nb-O system (e.g., Holmberg et al., 2004; Baca et al., 2005). The presence of up to 40% M2 phase in these catalysts was claimed to improve the acrylonitrile yield due to the ability of this phase to actively and selectively convert the propylene intermediate desorbed from the M1 phase into acrylonitrile. However, these studies were conducted at ≤ 50% propane conversion when significant propylene formation (10–30 mol.% selectivity) over the pure M1 phase catalyst was observed, whereas the selectivity to propylene over the pure M1 phase was typically < 4 mol.% at high propane conversion > 60% (e.g., Grasselli, 2005; Watanabe and Ueda, 2006).

In order to elucidate the role of the M1/M2 phase cooperation, propane ammoxidation was conducted over (1) an original $Mo_1V_{0.3}Te_{0.17}Nb_{0.12}$ catalyst containing 75% M1 and 25% M2 and (2) a pure M1 phase obtained from the original catalyst by selective dissolution of the M2 phase in H_2O_2 (Korovchenko et al., 2008). A subsequent elemental analysis of this H_2O_2 solution indicated that the M2 phase formed *in situ* together with the M1 phase in the Nb-containing $Mo_1V_{0.3}Te_{0.17}Nb_{0.12}$ system was essentially Nb-free in agreement with previous observations (e.g., Grasselli, 2005). The results of this study conducted at an optimal temperature of 693 K (Figure 7.14) confirmed that the M1 + M2 mixture was more efficient in propane ammoxidation at low to moderate propane conversion (< 50%) because the Nb-free M2 phase was much more active than the Nb-containing M1 phase in ammoxidation of the propylene intermediate formed over the M1 phase, e.g., the initial rates of ~5 and 0.6 μmol/m_2·s, respectively, at 693 K in $C_3H_6/NH_3/O_2/He$ = 6/7/17/70. However, this advantage was lost at longer reactor residence times and higher propane conversion as the M1 phase became more efficient in propylene ammoxidation. Therefore, at high propane conversions important for applied reasons, the M1 phase is *the only crystalline phase required*

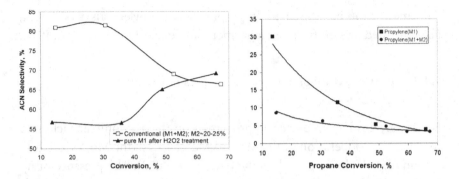

Fig. 7.14. Selectivity to acrylonitrile, ACN (*left*) and propylene (*right*) for propane ammoxidation at 693 K over (M1 + M2) and pure M1 Mo-V-Te-Nb-O catalysts. W_{cat} = 0.2 g (diluted with 0.5 g SiC); F = 20 ml/min; feed composition: $C_3H_8/NH_3/O_2/He$ = 6/7/17/70 (Figure 7 from Korovchenko *et al.*, 2008).

for the activity and selectivity of the bulk Mo-V-Te-Nb-O catalysts in propane ammoxidation to acrylonitrile.

7.4.5 Surface termination of M1 phase

Recently, the Schlögl group at the Fritz Haber Institute in Berlin (see Site Reports-Europe on the International Assessment of Research in Catalysis by Nanostructured Materials website, www.wtec.org/private/catalysis; password required, contact WTEC) reported that the *ab* planes in the Mo-V-Te-Nb-O M1 phase terminated with a nanometer-thick disordered layer (Wagner *et al.*, 2006). They concluded that it corresponded to nanoparticles supported on the M1 phase made from a supramolecular network of oxoclusters and containing the active and selective surface sites for propane (amm)oxidation. However, the results of recent high-resolution TEM and low-energy ion scattering (LEIS) studies of the *ab* plane in the M1 catalyst (Korovchenko *et al.*, 2008; Guliants *et al.*, 2005a) provided compelling evidence that the active and selective sites may be associated with a specific surface termination of the M1 phase rather than supramolecular structures (Wagner *et al.*, 2006).

As the LEIS data shown in Figure 7.15 suggest, only the topmost surface displayed the chemical composition different from that in the bulk. Moreover, a study of propane oxidation to acrylic acid over the Mo-V-Te-O

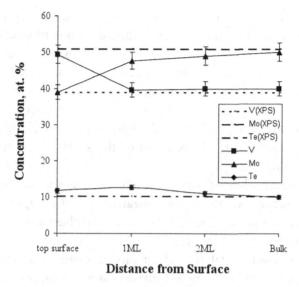

Fig. 7.15. Mo, V, and Te concentration profiles in the surface region of the $Mo_{0.3}V_{0.6}Te_{0.1}O$ M1 catalyst (Figure 7 from Guliants et al., 2005a).

M1 phase catalysts indicated that the rates of propane consumption and formation of propylene and acrylic acid depended on the topmost surface V concentration, whereas no dependence of these reaction rates on either the surface Mo or Te concentrations was observed (Guliants et al., 2005a). Average reaction orders of 2.3 ± 0.4, 1.6 ± 0.6, and 3.3 ± 0.6 were found, respectively, for propane consumption, propylene, and acrylic acid formation at 360–450°C with respect to the topmost surface V concentration, suggesting that multiple surface VO_x sites were more efficient in propane activation and formation of acrylic acid over the model Mo-V-Te-O catalysts than the isolated VO_x sites. These results suggest that the surface VO_x is the active and selective species, not only in the initial step of propane activation and ODH to propylene, but also in the subsequent steps of the allylic oxidation of propylene intermediate to acrolein and acrylic acid.

These recent findings suggest further that the M1 phase is the only crystalline phase required for the activity and selectivity of the Mo-V-Te-Nb-O catalysts in propane ammoxidation to acrylonitrile at high propane conversion, whereas a cooperation between the M1 and M2 phases is observed at low propane conversion. High-resolution TEM and LEIS studies indicated that the M1 phase is terminated by a *surface monolayer* proposed to be responsible for its activity and selectivity in propane (amm)oxidation. The experimental evidence is mounting that selective propane (amm)oxidation occurs at the surface ab planes of the M1 phase. Multiple VO_x sites present in the surface ab planes of the M1 phase were proposed to be highly efficient for this 8-ē oxidation reaction, whereas other surface MO_x species tune the activity and selectivity of the surface VO_x species by forming bridging V-O-M bonds and modifying the surface acidity to control the residence times of the surface intermediates and reaction products.

From a practical viewpoint, the results of these studies strongly indicate that further synthetic optimization of the topmost surface composition and surface promotion of the M1 phase with metal oxides species to form new surface active V-O-M sites are two highly promising directions to enhance catalytic performance of these mixed metal oxides in propane (amm)oxidation. The M1 phase possesses unique catalytic properties among mixed metal oxides, because it is capable of selectively catalyzing three alkane transformations reactions, namely propane ammoxidation to acrylonitrile, propane oxidation to acrylic acid, and oxidative dehydrogenation of ethane. Therefore, improved understanding of the surface structure–reactivity relationships for this unique model system offers a possibility of not only molecular engineering of such mixed metal oxide catalysts for propane (amm)oxidation, but also fundamentally advancing the field of selective alkane (amm)oxidation over bulk mixed metal oxides.

7.5 Future Trends

The foregoing review provided recent examples of size-, shape-, structure- and composition-dependent behavior of catalyst nanoparticles employed in alkylation, dehydrogenation, hydrogenation, and selective oxidation reactions of fossil resources to chemicals. Innovation in these areas is largely driven by novel synthesis of (nano)porous and nanostructured catalytic materials.

Several new classes of porous alkylation catalysts have recently emerged: periodic mesoporous organosilicas (PMOs), nanosized zeolites, hierarchical mesoporous zeolites, ultralarge-pore open (super)tetrahedral frameworks, metal organic frameworks (MOF), and amorphous microporous molecular sieves. The primary objective in the synthesis of such porous materials is to obtain stable ultralarge-pore frameworks with desirable acidity, shape selectivity, and improved mass transport characteristics for large organic molecules. Whereas supramolecular templating approaches were successful in introducing mesoporosity into known zeolite structures, the discovery of novel ultralarge-pore frameworks with desirable acidity remains largely a serendipitous process, due to the complexity of zeolite crystallization phenomena in the presence of organic structure-directing agents.

Several theoretical approaches to predicting structure-directing agents on the basis of a geometric and energetic fit to the zeolitic cavity including *de novo* design methods have met with some initial success (Catlow et al., 2003; Lewis, 2004). However, these expectations have been tempered by new knowledge that formation of stable zeolite frameworks is ultimately governed on a longer length scale by aggregation of nanosized building units during synthesis (Ramanan et al., 2004; Kremer et al., 2005; Magusin et al., 2005; Rimer et al., 2006). Therefore, accurate description of nucleation and growth of open frameworks on multiple length scales (from a few Å to several nm) and time scales (up to several hours) is critical to being able to design novel open frameworks. On the other hand, from practical considerations it is highly desirable to predict framework locations of catalytically important heteroatoms, as well as catalytic behavior and stability of new porous structures under relevant experimental conditions. Recent studies show that this work is currently underway, and some first promising results are emerging (e.g., Jorge et al., 2005; 2006; Auerbach et al., 2006).

Noble metal nanoparticles (Pt, Pd, Rh, etc.) have been extensively employed to catalyze a wide range of dehydrogenation, hydrogenation, and selective oxidation reactions of organic molecules. In recent years, nanosized Au and its alloys with other metals, e.g., Pd, have shown tremendous promise as superior catalysts for a wide range of selective transformations. Noble metal nanoparticles have been synthesized by a variety of soft chemistry and vapor-phase methods and deposited on solid supports, both inorganic and polymeric, or stabilized by dendritic polymers in solution. Despite an explosion of research on nanoparticles of Au and other noble metals, fundamental understanding of unique size-dependent catalytic behavior of pure gold and its alloys with other metals, as well as the role of support, is still rather limited.

One reason for this slow progress is that the most interesting Au nanoparticles from the catalysis viewpoint contain several hundred Au atoms, and due to their size, such discrete systems are not fully amenable to accurate but computationally expensive high-level *ab initio* molecular electronic structure methods. Although periodic *ab initio* methods are more expedient, the model of a periodic metallic Au is no longer an accurate representation for catalytically relevant Au nanoparticles in the size range 1.5–3 nm. On the other hand, experimental description of catalytic Au nanoparticles is also limited, due to polydispersity of Au nanoparticle populations and the difficulty of describing the crystal shape of such small particles. How would one begin to quantify the catalytic roles of various crystal faces, edges, steps, and other defects that may exist on surfaces of such small Au crystallites? Another challenge inherent to such small nanoparticles is their limited stability and rapid coarsening, even under mild conditions of liquid phase catalytic processes at room temperature. Other obstacles that impede both fundamental and practical advances include the poisoning and leaching that these catalysts experience under liquid-phase and solvent-free reaction conditions. Therefore, development of novel approaches to synthesize and characterize stable metal nanoparticles with tightly controlled sizes in the 1–4 nm range is critical to gaining fundamental understanding of such catalytic systems.

In this regard, the work of Üli Heiz at the Technical University of Munich (see Site Reports-Europe on the International Assessment of Research in Catalysis by Nanostructured Materials website, www.wtec.org/private/catalysis; password required, contact WTEC) is highly promising in terms of generating and soft-landing tightly size-controlled Au and other metal clusters containing up to 100 atoms on well-defined single-crystal surfaces of inorganic supports (Gilb *et al.*, 2006). Heiz's group, together with the Landman group at the Georgia Institute of Technology, showed a unique size-dependent catalytic behavior of very small Au_8 clusters in CO oxidation (Yoon *et al.*, 2005). Future advances will depend on being able to expand and tightly control the size of Au and other metal nanoparticles in the range of most interest for heterogeneous catalysis,

300–400 atoms, where the electronic structure is in transition regime from that of a large molecule to that of a metal.

Molecular imprinting approaches attempt to mimic nature in designing heterogeneous catalysts that, akin to enzymes, display catalytic cavity shape complementary to that of reactants or even reaction intermediates. To date, these approaches have been based on the "lock and key" concept of enzymatic catalysis originally proposed by Emil Fischer in 1894, which assumes a static character of an enzyme binding pocket (Fischer, 1894). However, while this model explains enzyme specificity, it fails to explain the stabilization of the transition state observed in enzymatic catalysis and has been replaced by a dynamic view of "induced fit" (Koshland, 1958), according to which both the enzyme active site cavity and the substrate undergo conformational changes. Despite the awareness of the importance of dynamic behavior of enzymes, a unified description of enzymatic catalysis is still incomplete, and a number of proposals are being discussed (e.g., Schramm, 2006). These proposals include *ground state destabilization*, which is accomplished by reducing the reaction entropy change by bringing substrates together in the correct orientation to react; and *stabilization of the transition state*, through lowering the activation energy by, for example, straining the shape of a substrate by binding the transition-state conformation of the substrate/product molecules, so that the enzyme distorts the bound substrate(s) into their transition state form, thereby reducing the amount of energy required to complete the transition. Recent studies have provided new insights into the connection between internal dynamics of enzymes and their mechanism of catalysis via so-called "rate-promoting vibrations", which help facilitate catalysis due to atomic motion within the protein along the reaction coordinate (e.g., Eisenmesser *et al.*, 2005). Therefore, future advances in the molecular design of highly active and selective enzyme mimics are likely to depend on our ability to capture both static and dynamic characteristics of enzyme binding pockets in molecularly imprinted catalysts.

The bulk mixed metal oxides of vanadium, molybdenum, and other transition metals have shown great promise as highly active and selective oxidation catalysts. Despite unique ability of some of these crystalline structures to selectively oxidize small alkanes into more reactive and valuable chemical intermediates, such as the M1 phase for propane ammoxidation to acrylonitrile, fundamental understanding of surface molecular structure-reactivity relationships remains highly limited, due to the structural and compositional complexity of these polycrystalline mixed oxides as well as to the limitations of current experimental methods of surface characterization of such systems under any conditions (ultrahigh-vacuum, *in situ*, during catalysis at high pressure). In this regard, it may be argued that our understanding of these complex oxides, from both the experimental and theoretical perspectives, lags behind our knowledge of better defined zeolite and metal catalyst systems. Therefore, significant opportunities exist to advance fundamental understanding of such systems with the ultimate goal of designing improved catalysts for selective alkane oxidation.

Recent studies have shown that such fundamental advances in understanding of mixed metal oxide catalysts may be possible through combined experimental and theoretical approaches. Improved knowledge of topmost surface compositions in these complex oxides may come from LEIS and high-pressure XPS studies (Bluhm *et al.*, 2007). However, these experimental techniques provide compositional information that is averaged over all crystal planes present in typically polycrystalline mixed metal oxides. Recent studies have provided some indications that these limitations may be overcome via careful passivation of the catalytic surface, followed by selective exposure of specific crystal planes by simple mechanical crushing of a passivated sample (Trunschke *et al.*, 2006). However, these approaches will only provide the chemical composition of the active surface. New insights into the surface structure in these materials may come from novel *in situ* X-ray absorption techniques (Havecker *et al.*, 2004) and further development of combined chemical probe chemisorption/LEIS methods (Guliants *et al.*, 2006) to study the surface reactivity of various metal oxide sites as well as their spatial arrangement at the surface. Ultimately, theoretical approaches, i.e., quantum theory, guided by experimental insights, may assist in judicious selection of first simple cluster models of the active and selective surface sites present in bulk mixed metal oxides. As we grow confident in these early models, these first steps should be expanded to employ larger and more realistic periodic slab models of the catalytic surfaces and techniques that will enable modeling the reactivity under realistic conditions, e.g., temperatures, pressures, and times (Reuter *et al.*, 2005). Ultimately, the goal of these studies is to develop our understanding of these complex systems to a point where one is able to accurately predict catalytic behavior of surface structures stabilized by the mixed metal oxide bulk under realistic conditions.

References

Abad, A., Almela, C., Corma, A., and García, H. (2006). Efficient chemoselective alcohol oxidation using oxygen as oxidant. Superior performance of gold over palladium catalysts, *Tetrahedron*, 62, pp. 6666–6672.

Anastas, P. T., and Warner, J. C. (1998). *Green Chemistry: Theory and Practice* (Oxford University Press, Oxford).

Andres, R., De Jesus, E., and Flores, J. C. (2007). Catalysts based on palladium dendrimers, *New J. Chem.*, 31, pp. 1161–1191.

Asefa, T., MacLachlan, M. J., Coombs, N., and Ozin, G. A. (1999). Periodic mesoporous organosilicas with organic groups inside the channel walls, *Nature*, 402, pp. 867–871.

Astruc, D., Lu, F., and Aranzaes, J. R. (2005). Nanoparticles as recyclable catalysts. The frontier between homogeneous and heterogeneous catalysis, *Angew. Chem., Int. Ed.*, 44, pp. 7852–7872.

Auerbach, S. M., Ford, M. H., and Monson, P. A. (2005). New insights into zeolite formation from molecular modeling, *Curr. Opin. Colloid Interface Sci.*, 10, pp. 220–225.

Baca, M., Aouine, M., Dubois, J. L., and Millet, J. M. M. (2005). Synergetic effect between phases in MoVTe(Sb)NbO catalysts used for the oxidation of propane into acrylic acid, *J. Catal.*, 233, pp. 234–241.

Bartholomew, C. H., and Farrauto, R. J. (2005). *Fundamentals of Industrial Catalytic Processes*, 2nd Ed. (Wiley–AIChE).

Bell, A. T. (2003). The impact of nanoscience on heterogeneous catalysis, *Science*, 299, pp. 1688–1691.

Berhault, G., Bisson, L., Thomazeau, C., Verdon, C., and Uzio, D. (2007). Preparation of nanostructured Pd particles using a seeding synthesis approach: Application to the selective hydrogenation of buta-1,3-diene, *Appl. Catal. A*, 327, pp. 32–43.

Biella, S., Castiglioni, G. L., Fumagalli, C., Prati, L., and Rossi, M. (2002). Application of gold catalysts to selective liquid phase oxidation, *Catal. Today*, 72, pp. 43–49.

Bluhm, H., Havecker, M., Knop-Gericke, A., Kishinova, M., Schlögl, R., and Salmeron, M. (2007). In situ X-ray photoelectron spectroscopy studies of gas-solid interfaces at near-ambient conditions, *MRS Bull.*, 32, pp. 1022–1030.

Bond, G. C. (2005). *Metal-Catalysed Reactions of Hydrocarbons* (Springer, New York) (and references therein).

Bond, G. C. (2007). The electronic structure of platinum-gold alloy particles, *Plat. Met. Rev.*, 51, pp. 63–68.

Bond, G. C., Louis, C., and Thompson, D. T. (2006). *Catalysis by Gold*, Catalytic Science Series, Vol. 6. (Imperial College Press, London).

Bowker, M. (1998). *The Basis and Applications of Heterogeneous Catalysis* (Oxford Science Publications).

Burato, C., Centomo, P., Pace, G., Favaro, M., Prati, L., and Corain, B. (2005). Generation of size-controlled palladium(0) and gold(0) nanoclusters inside the nanoporous domains of gel-type functional resins, *J. Mol. Catal. A*, 238, pp. 26–34.

Catlow, C. R. A., Coombes, D. S., Slater, B., Lewis, D. W., and Pereira, J. C. G. (2003). Modeling nucleation and growth in zeolites, in *Handbook of Zeolite Science and Technology*, eds. Auerbach, S. M., Carrado, K. A., and Dutta, P. K. (Marcel Dekker, Inc, New York), pp. 91–128.

Chandler, B. D., and Gilbertson, J. D. (2006). Dendrimer-encapsulated bimetallic nanoparticles: Synthesis, characterization, and applications to homogeneous and heterogeneous catalysis, *Top. Organ. Chem.*, 20, pp. 97–120.

Chen, A., Zhang, W., Li, X., Tan, D., Han, X., and Bao, X. (2007). One-pot encapsulation of Pt nanoparticles into the mesochannels of SBA-15 and their catalytic dehydrogenation of methylcyclohexane, *Catal. Lett.*, 119, pp. 159–164.

Comotti, M., Della Pina, C., Matarrese, R., Rossi, M., and Siani, A. (2005). Oxidation of alcohols and sugars using Au/C catalysts, *Appl. Catal. A*, 291, pp. 204–209.

Corma, A. (2003). State of the art and future challenges of zeolites as catalysts, *J. Catal.*, 216, pp. 298–312.

Corma, A., and Díaz-Cabañaz, M. J. (2006). Amorphous microporous molecular sieves with different pore dimensions and topologies: Synthesis, characterization and catalytic activity, *Micropor. Mesopor. Mater.*, 89, pp. 39–46.

Corma, A., and Serna, P. (2006). Chemoselective hydrogenation of nitro compounds with supported gold catalysts, *Science*, 313, pp. 332–334.

Council for Chemical Research (1998). *Vision 2020 catalysis report*, www.ccrhq.org/vision/index/roadmaps/catrep.html.

Davis, M. E. (2002). Ordered porous materials for emerging applications, *Nature*, 417, pp. 813–821.

DeSanto, P., Jr., Buttrey, D. J., Grasselli, R. K., Lugmair, C. G., Volpe, A. F., Toby, B. H., and Vogt, T. (2004). Structural aspects of the M1 and M2 phases in MoVNbTeO propane ammoxidation catalysts, *Z. Kristallogr.*, 219, pp. 152–165.

Edwards, J. K., Solsona, B. E., Landon, P., Carley, A. F., Herzing, A., Kiely, C. J., and Hutchings, G. J. (2005). Direct synthesis of hydrogen peroxide from H_2 and O_2 using TiO_2-supported Au-Pd catalysts, *J. Catal.*, 236, pp. 69–79.

Eisenmesser, E. Z., Millet, O., Labeikovsky, W., Korzhnev, D. M., Wolf-Watz, M., Bosco, D. A., Skalicky, J. J., Kay, L. E., and Kern, D. (2005). Intrinsic dynamics of an enzyme underlies catalysis, *Nature*, 438, pp. 117–121.

Ertl, G., Knözinger, H., and Weitkamp, J., eds. (1997). *Handbook of Heterogeneous Catalysis*, Vol. 5 (VCH, Weinheim).

Favier, I., Gómez, M., Muller, G., Picurelli, D., Nowicki, A., Roucoux, A., and Bou, J. (2007). Synthesis of new functionalized polymers and their use as stabilizers of Pd, Pt, and Rh nanoparticles. Preliminary catalytic studies, *J. Appl. Polymer Sci.*, 105, pp. 2772–2782.

Fischer, E. (1894). Einfluss der configuration auf die wirkung der enzyme, *Ber. Dt. Chem. Ges.*, 27, pp. 2985–2993.

Fukuoka, A., Higashimoto, N., Sakamoto, Y., Inagaki, S., Fukushima, Y., and Ichikawa, M. (2001). Preparation and catalysis of Pt and Rh nanowires and particles in FSM-16, *Micropor. Mesopor. Mater.*, 48, pp. 171–179.

Gates, B. C. (1992). *Catalytic Chemistry* (Wiley, New York City).

Gilb, S., Arenz, M., and Heiz, Ü. (2006). Monodispersed cluster-assembled materials, *Mater. Today*, 9, pp. 48–49.

Grasselli, R. K. (2002). Fundamental principles of selective heterogeneous oxidation catalysis, *Top. Catal.*, 21, pp. 79–88.

Grasselli, R. K. (2005). Selectivity issues in (amm)oxidation catalysis, *Catal. Today*, 99, pp. 23–31 (and references therein).

Guan, Y., Li, Y., Santen, R. A., Hensen, E. J. M., and Li, C. (2007). Controlling reaction pathways for alcohol dehydration and dehydrogenation over FeSBA-15 catalysts, *Catal. Lett.*, 117, pp. 18–24.

Guliants, V. V., Bhandari, R., Brongersma, H. H., Knoester, A., Gaffney, A. M., and Han, S. (2005a). A study of the surface region of the Mo-V-Te-O catalysts for propane oxidation to acrylic acid, *J. Phys. Chem. B*, 109, pp. 10234–10242.

Guliants, V. V., Bhandari, R., Swaminathan, B., Vasudevan, V. K., Brongersma, H. H., Knoester, A., Gaffney, A. M., and Han, S. (2005b). Roles of surface Te, Nb, and Sb oxides in propane oxidation to acrylic acid over bulk orthorhombic Mo-V-O phase, *J. Phys. Chem. B*, 109, pp. 24046–24055.

Guliants, V. V., Bhandari, R., Hughett, A. R., Bhatt, S., Schuler, B. D., Brongersma, H. H., Knoester, A., Gaffney, A. M., and Han, S. (2006). Probe molecule chemisorption-low energy ion scattering study of surface active sites present in the orthorhombic Mo-V-(Te-Nb)-O catalysts for propane (amm)oxidation, *J. Phys. Chem. B*, 110, pp. 6129–6140.

Guo, X. C., and Madix, R. J. (1995). Selective hydrogenation and H-D exchange of unsaturated hydrocarbons on Pd(100)-p(1x1)-H(D), *J. Catal.*, 155, pp. 336–344.

Haruta, M. (2003). When gold is not noble: Catalysis by nanoparticles, *Chem. Rec.*, 3, pp. 75–87.

Hashmi, A. S. K., and Hutchings, G. J. (2006). Gold catalysis, *Angew. Chem. Int. Ed.*, 45, pp. 7896–7936.

Havecker, M., Knop-Gericke, A., Bluhm, H., Kleimenov, E., Mayer, R. W., Fait, M., and Schlögl, R. (2004). Dynamic surface behaviour of VPO catalysts under reactive and non-reactive gas compositions: An in situ XAS study, *Appl. Surf. Sci.*, 230, pp. 272–282.

Hermes, S., Schröter, M.-K., Schmid, R., Khodeir, L., Muhler, M., Tissler, A., Fischer, R. W., and Fischer, R. A. (2005). Metal@MOF: Loading of highly porous coordination polymers host lattices by metal organic chemical vapor deposition, *Angew. Chem. Int. Ed.*, 44, pp. 6237–6241.

Holmberg, J., Grasselli, R. K., and Andersson, A. (2004). Catalytic behavior of M1, M2, and M1/M2 physical mixtures of the Mo-V-Nb-Te-oxide system in propane and propene ammoxidation, *Appl. Catal. A*, 270, pp. 121–134.

Horiuchi, J., and Polanyi, M. (1934). Exchange reactions of hydrogen on metallic catalysts, *Trans. Faraday Soc.*, 30, pp. 1164–1172.

Hughes, M. D., Xu, Y.-J., Jenkins, P., McMorn, P., Landon, P., Enache, D. I., Carley, A. F., Attard, G. A., Hutchings, G. J., King, F., Stitt, E. H., Johnston, P., Griffin, K., and Kiely, C. J. (2005). Tunable gold catalysts for selective hydrocarbon oxidation under mild conditions, *Nature*, 437, pp. 1132–1135.

Ishida, T., and Haruta, M. (2007). Gold catalysts: Towards sustainable chemistry, *Angew. Chem. Int. Ed.*, 46, pp. 7154–7156.

Ishida, T., Okamoto, S., Makiyama, R., and Haruta, M. (2008). Aerobic oxidation of glucose, 1-phenylethanol over gold nanoparticles directly deposited on ion-exchange resins, *Appl. Catal. A: Gen.*, DOI:10.1016/j.apcata.2008.10.049.

Jorge, M., Auerbach, S. M., and Monson, P. A. (2005). Modeling spontaneous formation of precursor nanoparticles in clear-solution zeolite synthesis, *J. Am. Chem. Soc.*, 127, pp. 14388–14400.

Jorge, M., Auerbach, S. M., and Monson, P. A. (2006). Modeling the thermal stability of precursor nanoparticles in zeolite synthesis, *Mol. Phys.*, 104, pp. 3513–3522.

Katano, S., Kato, H. S., Kawai, M., and Domen, K. (2003). Selective partial hydrogenation of 1,3-butadiene to butene on Pd(110): Specification of reactant adsorption states and product stability, *J. Phys. Chem. B*, 107, pp. 3671–3674.

Katz, A., and Davis, M. E. (2000). Molecular imprinting of bulk, microporous silica, *Nature*, 403, pp. 286–289.

Ketchie, W. C., Murayama, M., and Davis, R. J. (2007). Selective oxidation of glycerol over carbon-supported AuPd catalysts, *J. Catal.*, 250, pp. 264–273.

Kónya, Z., Puntes, V. F., Kiricsi, I., Zhu, J., Ager, J. W., Ko, M. K., Frei, H., Alivisatos, P., and Somorjai, G. A. (2003). Synthetic insertion of gold nanoparticles into mesoporous silica, *Chem. Mater.*, 15, pp. 1242–1248.

Kónya, Z., Puntes, V. F., Kiricsi, I., Zhu, J., Alivisatos, P., and Somorjai, G. A. (2002). Novel two-step synthesis of controlled size and shape platinum nanoparticles encapsulated in mesoporous silica, *Catal. Lett.*, 81(3–4), pp. 137–140.

Korovchenko, P., Shiju, N. R., Dozier, A. K., Graham, U. M., Guerrero-Pérez, M. O., and Guliants, V. V. (2008). M1 to M2 phase transformation and phase cooperation in bulk mixed metal Mo-V-M-O (M = Te, Nb) catalysts for selective ammoxidation of propane, *Top. Catal.*, DOI: 10.1007/s11244-008-9098-8, published online 10 June 2008.

Koshland, D. E. (1958). Application of a theory of enzyme specificity to protein synthesis, *Proc. Natl. Acad. Sci.*, 44, pp. 98–104.

Kremer, S. P. B., Kirschhock, C. E. A., Jacobs, P. A., and Martens, J. A. (2005). Synthesis and characterization of zeogrid molecular sieves, *Comptes Rendus Chimie*, 8, pp. 379–390.

Lewis, D. W. (2004). Template-host interaction and template design, in *Computer Modelling of Microporous Materials*, eds. Catlow, C. R. W., van Santen, R. A., and Smit, B. (Academic Press), pp. 243–265.

Li, Y., Xia, H., Fan, F., Feng, Z., van Santen, R. A., Hensen, E. J. M., and Li, C. (2008). Iron-functionalized Al-SBA-15 for benzene hydroxylation, *Chem. Commun.*, pp. 774–776.

Li, Y., Feng, Z., Xin, H., Fan, F., Zhang, J., Magusin, P. C. M. M., Hensen, E. J. M., van Santen, R. A., Yang, Q., and Li, C. (2006). Effect of aluminum on the nature of the iron species in Fe- SBA-15, *J. Phys. Chem. B*, 110, pp. 26114–26121.

Magusin, P. C. M. M., Zorin, V. E., Aerts, A., Houssin, C. J. Y., Yakovlev, A. L., Kirschhock, C. E. A., Martens, J. A., and van Santen, R. A. (2005). Template-aluminosilicate structures at the early stages of zeolite ZSM-5 formation. A combined preparative, solid-state NMR, and computational study, *J. Phys. Chem. B*, 109, pp. 22767–22774.

McCrea, K. R., and Somorjai, G. A. (2000). FG-surface vibrational spectroscopy studies of structure sensitivity and insensitivity in catalytic reactions: Cyclohexene dehydrogenation and ethylene hydrogenation on Pt (111) and Pt (100) crystal surfaces, *J. Mol. Catal. A: Chem.*, 163, pp. 43–53.

Mertens, P. G. N., Poelman, H., Yec, X., Vankelecom, I. F. J., Jacobs, P. A., and De Vos, D. E. (2007). Au^0 nanocolloids as recyclable quasihomogeneous metal catalysts in the chemoselective hydrogenation of α, β-unsaturated aldehydes and ketones to allylic alcohols, *Catal. Today*, 122, pp. 352–360.

Miyamura, H., Matsubara, R., Miyazaki, Y., and Kobayashi, S. (2007). Aerobic oxidation of alcohols at room temperature and atmospheric conditions catalyzed by reusable gold nanoclusters stabilized by the benzene rings of polystyrene derivatives, *Angew. Chem. Int. Ed.*, 46, p. 4151.

Morbidelli, M., Gavrillidis, A., and Varma, A. (2001). *Catalyst Design: Optimal Distribution of Catalyst in Pellets, Reactors, and Membranes* (Cambridge University Press, New York City).

Mori, K., Hara, T., Mizugaki, T., Ebitani, K., and Kaneda, K. (2004). Hydroxyapatite-supported palladium nanoclusters: A highly active heterogeneous catalyst for selective oxidation of alcohols by use of molecular oxygen, *J. Am. Chem. Soc.*, 126, pp. 10657–10666.

Narayanan, R., and El-Sayed, M. A. (2004a). Shape-dependent catalytic activity of platinum nanoparticles in colloidal solution, *Nano Lett.*, 4, pp. 1343–1348.

Narayanan, R., and El-Sayed, M. A. (2004b). Effect of nanocatalysis in colloidal solution on the tetrahedral and cubic nanoparticle shape: Electron-transfer reaction catalyzed by platinum nanoparticles, *J. Phys. Chem. B*, 108, pp. 5726–5733.

Park, K. H., Jang, K., Kim, H. J., and Son, S. U. (2007). Near-monodisperse tetrahedral rhodium nanoparticles on charcoal: The shape-dependent catalytic hydrogenation of arenes, *Angew. Chem. Int. Ed.*, 46, pp. 1152–1155.

Pârvulescu, V. I., Pârvulescu, V., Endruschat, U., Filoti, G., Wagner, F. E., Kübel, C., and Richards, R. (2006). Characterization and catalytic-hydrogenation behavior of SiO_2 embedded nanoscopic Pd, Au, and Pd–Au alloy colloids, *Chem. Eur. J.*, 12, pp. 2343–2357.

Prati, L., and Rossi, M. (1998). Gold on carbon as a new catalyst for selective liquid phase oxidation of diols, *J. Catal.*, 176, pp. 552–560.

Pyrz, W. D., Blom, D. A., Shiju, N. R., Guliants, V. V., Vogt, T., and Buttrey, D. J. (2008). Using aberration-corrected STEM imaging to explore chemical and structural variations in the M1 Phase of the MoVNbTeO oxidation catalyst, *J. Phys. Chem. C*, 112(27), pp. 10043–10049.

Ramanan, H., Kokkoli, E., and Tsapatsis, M. (2004). On the TEM and AFM evidence of zeosil nanoslabs present during the synthesis of silicalite-1 (Comment), *Angew. Chem. Int. Ed.*, 43, pp. 4558–4561.

Reuter, K., Stampfl, C., and Scheffler, M. (2005). Ab initio atomistic thermodynamics and statistical mechanics of surface properties and functions, in *Handbook of Materials Modeling, Pt., A.*, ed. Yip, S. (Springer), pp. 149–194.

Rimer, J. D., Fedeyko, J. M., Vlachos, D. G., and Lobo, R. F. (2006). Silica self-assembly and synthesis of microporous and mesoporous silicates, *Chem. Eur. J.*, 12, pp. 2926–2934.

Rupprechter, G. (2007). A surface science approach to ambient pressure catalytic reactions, *Catal. Today*, pp. 1263–1317.

Sakamoto, Y., Fukuoka, A., Higuchi, T., Shimomura, N., Inagaki, S., and Ichikawa, M. (2004). Synthesis of platinum nanowires in organic-inorganic mesoporous silica templates by photoreduction: Formation mechanism and isolation, *J. Phys. Chem. B*, 108, pp. 853–858.

Schoeman, B. J., Sterte, J., and Otterstedt, J. E. (1994). Colloidal zeolite suspensions, *Zeolites*, 14, pp. 110–116.

Schramm, V. L., ed. (2006). Principles of enzymatic catalysis. Special Issue, *Chem. Rev.*, 106, pp. 3029–3496.

Shiju, N. R., and Guliants, V. V. (2007). Microwave-assisted hydrothermal synthesis of monophasic Mo-V-Te-Nb-O mixed oxide catalyst for the selective ammoxidation of propane, *Chem. Phys. Chem.*, 8, p. 1615.

Silvestre-Alberto, J., Rupprechter, G., and Freund, H. J. (2005). Atmospheric pressure studies of selective 1,3-butadiene hydrogenation on Pd single crystals: Effect of CO addition, *J. Catal.*, 235, pp. 52–59.

Tada, M., and Iwasawa, Y. (2006). Advanced chemical design with supported metal complexes for selective catalysis, *Chem. Commun.*, pp. 2833–2844 (and references therein).

Telkar, M. M., Rode, C. V., Chaudhari, R. V., Joshi, S. S., and Nalawade, A. M. (2004). Shape-controlled preparation and catalytic activity of metal nanoparticles for hydrogenation of 2-butyne-1,4-diol and styrene oxide, *Appl. Catal. A*, 273, pp. 11–19.

Thomas, J. M., and Thomas, W. J. (1997). *Principles and Practice of Heterogeneous Catalysis*, 2nd Ed. (Wiley — VCH, Weinheim).

Toshima, N., Shiraishi, Y., Teranishi, T., Miyake, M., Tominaga, T., Watanabe, H., Brijoux, W., Bönnemann, H., and Schmid, G. (2001). Various ligand-stabilized metal nanoclusters as homogeneous and heterogeneous catalysts in the liquid phase, *Appl. Organ. Chem.*, 15, pp. 178–196.

Trunschke, A., Schlögl, R., Guliants, V. V., Knoester, A., and Brongersma, H. (2006). Surface analysis of the ab-plane of MoVTeNbOx catalysts for propane (amm)oxidation by low energy ion scattering (LEIS), presented at the 232nd ACS National Meeting, San Francisco, CA.

Tsuji, Y., and Fujihara, T. (2007). Homogeneous nanosize palladium catalysts, *Inorg. Chem.*, 46, pp 1895–1902.

Tsunoyama, H., Sakurai, H., Negishi, Y., and Tsukuda, T. (2005). Size-specific catalytic activity of polymer-stabilized gold nanoclusters for aerobic alcohol oxidation in water, *J. Am. Chem. Soc.*, 127, pp. 9374–9375.

van Santen, R. A., and Neurock, M. (2006). *Molecular Heterogeneous Catalysis: A Conceptual and Computational Approach* (Wiley — VCH, Weinheim).

Wagner, J. B., Timpe, O., Hamid, F. A., Trunschke, A., Wild, U., Su, D. S., Widi, R. K., Hamid, S. B. A., and Schlögl, R. (2006). Surface texturing of Mo-V-Te-Nb-Ox selective oxidation catalysts, *Top. Catal.*, 38, pp. 51–58.

Wahab, M. A., and Ha, C. S. (2005). Ruthenium-functionalized hybrid periodic mesoporous organosilicas: synthesis and structural characterization, *J. Mater. Chem.*, 15, pp. 508–516.

Wan, Y., and Zhao, D. (2007). On the controllable soft-templating approach to mesoporous silicates, *Chem. Rev.*, 107, pp. 2821–2860.

Watanabe, N., and Ueda, W. (2006). Comparative study on the catalytic performance of single-phase Mo-V-O-based metal oxide catalysts in propane ammoxidation to acrylonitrile, *Ind. Eng. Chem. Res.*, 45, pp. 607–614.

Whitesides, G. M., and Crabtree, G. W. (2007). Don't forget long-term fundamental research in energy, *Science*, 315, pp. 796–798.

Wilson, O. M., Knecht, M. R., Garcia-Martinez, J. C., and Crooks, R. M. (2006). Effect of Pd nanoparticle size on the catalytic hydrogenation of allyl alcohol, *J. Am. Chem. Soc.*, 128, pp. 4510–4511 (and references therein).

Xia, Y., Wang, W., and Mokaya, R. (2005). Bifunctional hybrid mesoporous organoaluminosilicates with molecularly ordered ethylene groups, *J. Am. Chem. Soc.*, 127, p. 790.

Xiao, F.-S., Wang, L., Yin, C., Lin, K., Di, Y., Li, J., Xu, R., Su, D. S., Schlögl, R., Yokoi, T., and Tatsumi, T. (2006). Catalytic properties of hierarchical mesoporous zeolites templated with a mixture of small organic ammonium salts and mesoscale cationic polymers, *Angew. Chem. Int. Ed.*, 45, pp. 3090–3093.

Yang, Q., Yang, J., Feng, Z., and Li, Y. (2005). Aluminum-containing mesoporous benzene-silicas with crystal-like pore wall structure, *J. Mater. Chem.*, 15, pp. 4268–4274.

Yang, Q., Li, Y., Zhang, L., Yang, J., Liu, J., and Li, C. (2004). Hydrothermal stability and catalytic activity of aluminum-containing mesoporous ethane-silicas, *J. Phys. Chem. B*, 108, pp. 7934–7937.

Yoon, B., Haekkinen, H., Landman, U., Woerz, A. S., Antonietti, J.-M., Abbet, S., Judai, K., and Heiz, Ü. (2005). Charging effects on bonding and catalyzed oxidation of CO on Au8 clusters on MgO, *Science*, 307, pp. 403–407.

Zhang, L., Abbenhuis, H. C. L., Gerritsen, G., Bhriain, N. N., Magusin, P. C. M. M., Mezari, B., Han, W., van Santen, R. A., Yang, Q., and Li, C. (2007). An efficient hybrid, nanostructured, epoxidation catalyst: Titanium silsesquioxane-polystyrene copolymer supported on SBA-15, *Chem. Eur. J.*, 13, pp. 1210–1221.

Zhao, D., Huo, Q., Feng, J., Chmelka, B. F., and Stucky, G. D. (1998). Nonionic tri-block and star diblock copolymer and oligomeric surfactant syntheses of highly ordered, hydrothermally stable, mesoporous silica structures, *J. Am. Chem. Soc.*, 120, pp. 6024–6036.

Zheng, N., and Stucky, G. D. (2006). A general synthetic strategy for oxide-supported metal nanoparticle catalysts, *J. Am. Chem. Soc.*, 218, pp. 14278–14280.

8

Applications: Renewable Fuels and Chemicals

George Huber

8.1 Introduction

Rising fossil fuel prices, combined with widespread concerns about the environmental effects of fossil fuels, are causing researchers worldwide to develop new ways of producing renewable fuels and chemicals. The field of heterogeneous catalysis will be an integral part of the process; no other scientific discipline is as well positioned to make a long-term impact on our ability to efficiently use our renewable resources. Today, catalysis allows us to convert our raw fossil fuel resources to fuels and chemicals that provide us with inexpensive raw materials that drive our economy. In the future, catalysis will be used to efficiently convert renewable resources into the raw materials and liquid fuels that we need to maintain our high living standard.

On the survey the WTEC panel sent to its hosts prior to its visits, the first question asked: What are the major needs, opportunities, or directions in catalysis research over the next 10- and 20-year time frames? Almost all hosts answered that the major research directions in catalysis have to do with sustainability and conversion of renewable resources

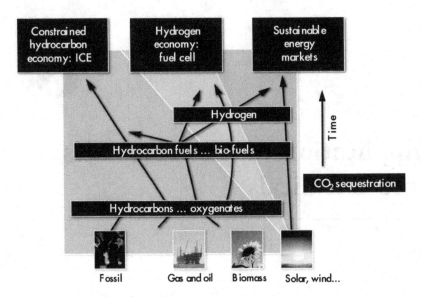

Fig. 8.1. The shift from a nonrenewable to a sustainable economy (courtesy of Shell Global Solutions BV).

(sunlight, biomass) to fuels and chemicals. This question was answered the same way in China, Japan, and in Europe by researchers in academia, industry, and in national labs. In addition, the WTEC panelists observed significant local media coverage related to renewable energy projects in the places we visited, indicating public interest in the work. Figure 8.1 shows how Shell envisions the transition to a sustainable energy economy. The field of catalysis will be integral to this transition, the timing of which depends on how quickly scientists and engineers are able to make key scientific and technical discoveries.

This chapter assesses the state-of-the-art of catalysis research and development in the United States, Europe, and Asia in the areas of renewable fuels and chemicals, first addressing biomass conversion and then solar energy conversion.

8.2 Key Observations

We are currently seeing an exponential growth in the area of catalysis related to biofuels and renewable energy in both industrial and academic

Applications: Renewable Fuels and Chemicals   241

institutions. This is likely to be a critical area for the United States in the next 10 to 20 years as we strive to reduce our dependence on imported oil and nonrenewable fossil energy resources generally. It should be emphasized that the purpose of this WTEC study was not to assess renewable energy, but to assess catalysis research. Therefore, this chapter should not be seen as a summary of worldwide renewable energy research, but as an insight into catalysis research that is being done in relationship to renewable energy.

This study makes six key observations related to catalysis applications to renewable fuels and chemicals:

- Catalysis is the key cross-cutting science that will allow efficient conversion of renewable resources (sunlight, biomass, hydrogen) into fuels and chemicals.
- Many research projects in catalytic biomass conversion are just now getting started; this new area appears to be growing exponentially.
- The United States is not devoting as much resources to catalysis and biofuel projects as many other parts of the world. Europe publishes the greatest number of papers related to biomass conversion with heterogeneous catalysis, especially related to biochemicals and biofuels. However, the papers published in the United States have a higher potential citation impact than those published in Europe.
- Land is available worldwide for renewable energy and biofuels. WTEC panelists were not able to visit a number of countries that are heavily invested in biofuels (Brazil, Canada, etc.).
- Japan is a world leader in photocatalytic splitting of water.
- China and India are rapidly developing industrial biofuel products (furfural, jatropha).

8.3 Biomass Conversion

8.3.1 *Biomass feedstocks*

Worldwide, there is a large amount of unused land that could be used to produce biomass, and there is already a large amount of biomass that is available for use as a feedstock to make liquid fuels and chemicals. Most of the existing biomass is in the form of lignocellulosic biomass (e.g., plants in the grass family, trees, and byproducts or waste products associated with their use). This form of biomass is not yet being used as a feedstock for liquid fuels because the technology has not yet been developed. Currently only oil-, sugar-, and starch-based crops are used commercially as biofuel feedstocks. The US Department of Agriculture (USDA) and Oak Ridge National Laboratory have estimated that the United States could sustainably produce 1.3 billion metric tons of dry biomass/year using its agricultural (72% of total) and forest (28% of total) resources and still meet its food, feed, and export demands (Perlack et al., 2005). This amount of biomass has the energy content of 3.8 billion boe (barrels of oil energy equivalent) (Klass, 1998). According to the US Energy Information Administration, the United States consumes over 7.5 bbl/yr or billion barrels of oil per year (EIA, 2008). According to the European Biomass Industry Association (EUBIA), Europe, Africa, and Latin America could produce biomass with an energy output of 1.4, 3.5, and 3.2 billion boe, respectively (e.g., see http://www.eubia.org/). The worldwide raw biomass energy potential in 2050 has been estimated to be 25 to 76 billion boe per year (IEA, 2004). In addition to contributing to energy and chemical production, cultivation of biomass feedstocks can also have a positive effect on agriculture; for example, the USDA has estimated that the net farm income in the United States could increase from US$3 to 6 billion annually if switchgrass became an energy crop (de la Torre Ugarte et al., 2003).

Governments around the world are promoting the biofuels industry by providing tax breaks, mandates, and agricultural subsidies, and devoting land to cultivation of biofuel feedstocks. The Chinese government is planning on devoting 50,000 square miles (approximately the size of Louisiana) for biofuels (Liu, 2007). India has identified 154,000 squares miles (approximately the size of California) of marginal land for biofuel production.

New crops are being developed for biofuel and energy usage. Jatropha, which is a warm weather oilseed crop that requires a maturation period of 5 years, can be grown on marginal land. Figure 8.2 shows a jatropha plantation in India and the seeds. The oil yield for a mature jatropha plant is 5–13 boe/ha-yr (ha = hectare). Jatropha plantations are being planted in many countries in Asia, Africa, and Central and South America. Members of the Indian government hope that biodiesel produced from jatropha will

Fig. 8.2. Jatropha plantation and oil seed (image sources (*left* to *right*): Jatropha World, http://www.jatrophabiodiesel.org/; Government of Kanpur Dehat, India, http://kanpurdehat.nic.in/photograph.htm; Greenfueltech, http://www.greenfueltech.net/).

Fig. 8.3. Miscanthus plantation and harvesting (image sources: (*left*) 1995 photo by Dr. I. Lewandowski, Institute for Crop Production and Grassland Research, University of Hohenheim, Germany, http://bioenergy.ornl.gov/papers/miscanthus/miscanthus.html; (*right*) University of Illinois at Urbana-Champaign, http://www.news.uiuc.edu/II/08/0207/index.html).

replace 20% of the national's diesel fuel by 2012; the government is providing loans (total amount US$34 million) to farmers to plant jatropha. Oil seeds are advantageous in that the technology for their conversion into biodiesel is inexpensive.

The yield of biomass from lignocellulosic energy crops can be up to 100 boe/ha-yr (Huber *et al.*, 2006). One example of a lignocellulosic energy crop that is being used in Europe is miscanthus, which is shown in Figure 8.3. Miscanthus has yield rates up to 75 boe/ha-yr and is a cold-tolerant perennial crop that requires very little fertilizer input. It is being grown commercially in Europe as a feed for electricity production in local area power stations. For miscanthus as for other biomass feed sources, the development of cost-effective and energy-efficient processes to transform the cellulose and lignin into fuels currently is hampered by a significant roadblock: a general lack of processes to deconstruct the biomass and transform the raw material into liquid fuels.

8.3.2 *Liquid fuels from biomass*

The main scientific challenge in converting biomass to a liquid fuel is how to efficiently remove oxygen from the biomass and produce a liquid product with good combustion and thermo-chemical properties (Huber *et al.*, 2006; Huber and Corma, 2007). Oxygen can be removed as CO, CO_2 and H_2O. The general stoichiometry for biomass-conversion to liquid fuels (with glucose as the feedstock) is shown in Equation (8.1). All biomass conversion reactions follow this stoichiometry. A range of liquid biofuels can be produced, including alkanes, aromatics, and oxygenated fuels. To maximize the amount of energy in the liquid fuel, the oxygen should be removed as a combination of CO_2 and H_2O. The ratio of oxygen removed as CO_2 versus H_2O depends on the products formed, as well as on whether or not hydrogen is used as a co-feed.

$$C_6H_{12}O_6 \rightarrow aC_xH_yO_z + bCO_2 + cCO + dH_2O. \qquad (8.1)$$

There are currently many different approaches being used to convert biomass to biofuels. A summary of these approaches is shown in Figure 8.4, which shows routes that use chemical catalysts in black, routes that use biological routes in green, and routes that use either biological or chemical catalysts in blue. As shown in Figure 8.4, most of the routes to make biofuels use chemical catalysts. During the WTEC visits in Europe and Asia, panelists observed that all of the routes in Figure 8.4 are being studied and that the optimal route to make biofuels is currently not known. The technologies shown in Figure 8.4 are in different stages of development. It should be emphasized that considerable research and development is required to make liquid fuels economically viable. Petroleum-based liquid fuels are produced in high-volume, very efficient processes that have been continually optimized over the last 80 years.

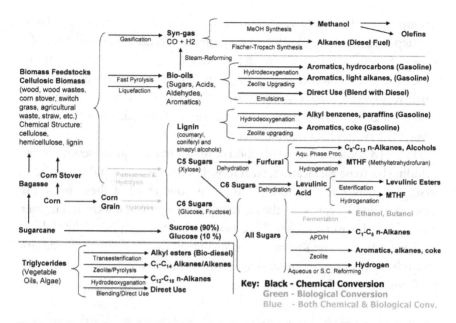

Fig. 8.4. Routes for production of liquid fuels from biomass (adapted from Huber *et al.*, 2006).

8.3.3 Biomass gasification and syngas conversion

Gasification of biomass and syngas conversion are probably the most technically developed areas of biofuel production. Several of the research groups the WTEC panel visited were studying biomass gasification, including Tsukuba University and Shell Global Solutions International. The panel also visited several groups that are studying syngas conversion, as described in Chapter 6.

There are several demonstration projects in Europe in which liquid fuels are being produced by the Fischer–Tropsch synthesis (FTS) of lignocellulosic biomass. One is a joint demonstration plant of Shell and the Energy Research Centre of the Netherlands (ECN) that ran for 1000 hours, in which liquid biofuel was produced from wood by gasification and FTS (Boerrigter, 2003; Boerrigter *et al.*, 2002). The researchers estimate the yield of diesel fuel from wood to be 120 L diesel fuel/metric ton biomass (Boerrigter, 2003). Scandinavia-based Neste Oil Corporation and Stora Enso Group (a forest products company) have signed an agreement to develop liquid biofuels by FTS (Renewable Energy Access, 2007). Their first step is to build an US$18.7 million demonstration plant for this technology. Several companies in the United States are also involved in demonstrating this technology. Range Fuels (http://www.rangefuels.com), which will receive US$76 million from the US Department of Energy, is building a 20-million-gallon-per-year plant in Georgia. Range Fuels will convert the syngas into ethanol instead of alkanes.

The FTS route to fuel production from biomass requires several steps, including biomass gasification, tar removal, water-gas shift, and syngas conversion to fuels. Catalysis research is being done in all of these areas. Tar removal probably represents the currently most significant technical barrier that needs to be overcome for this process to become commercially viable.

Researchers in the Kunimori–Tomishige group at Tsukuba University are studying the production of syngas from biomass and how to reduce the tar formation in this process. They have shown (see Figure 8.5) that

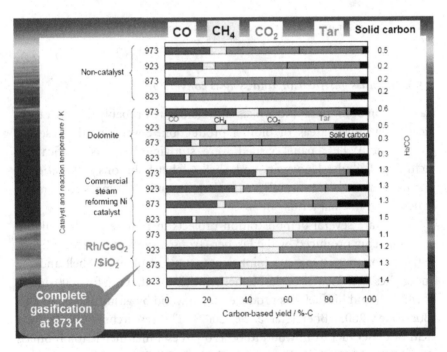

Fig. 8.5. Catalytic gasification of biomass (cedar) with air (adapted from Asadullah *et al.*, 2002). Catalysts for biomass-reforming to syngas. This figure shows that Rh/CeO$_2$/SiO$_2$ is able to efficiently reform the tars (courtesy of Dr. Keiichi Tomishige).

Rh/CeO$_2$/SiO$_2$ is particularly effective at converting biomass-derived tars into syngas (Asadullah *et al.*, 2002).

Several researchers in the United States, including the groups of Lanny Schmidt at the University of Minnesota (Salge *et al.*, 2006) and Umit Ozkan at Ohio State, and researchers at the National Renewable Energy Laboratory (NREL) are also studying methods for catalytic gasification. Another growing area of interest is syngas conversion into ethanol. This is an area that is being studied at NREL and Pacific Northwest National Laboratory (PNNL), as well as by Jerry Spivey at Louisiana State University, Jim Goodwin at Clemson University, and Bob Davis at the University of Virginia.

8.3.4 *Fast pyrolysis and bio-oil upgrading*

Fast pyrolysis of biomass is a technology for production of a liquid fuel called bio-oil. Fast pyrolysis of biomass involves rapid heating (i.e., 500°C/s) of the biomass to intermediate temperatures (i.e., 500°C) without air, followed by rapid cooling. The liquid products produced by fast pyrolysis are a nonequilibrium mixture of more than 300 different bio-oil compounds. Fast pyrolysis of cellulosic biomass has been shown to produce bio-oils that contain up to 70% of the energy content of biomass feed (Huber *et al.*, 2006; Mohan *et al.*, 2006; Bridgwater and Peacocke, 2000). Thus, fast pyrolysis is an efficient process for overcoming the recalcitrant nature of cellulosic biomass and producing liquid products.

The essential parameters for fast pyrolysis include the following (Bridgwater and Peacocke, 2000):

(1) A very high heating and heat transfer rate (i.e., 500°C/s).
(2) Carefully controlled temperatures around 450–550°C.
(3) Rapid cooling of the pyrolysis vapors (residence time of less than 1s).

One advantage of fast pyrolysis is that it can be done on a small scale (i.e., 50–100 tons biomass feed/day). This has led to the development of small-scale pyrolysis reactors like the one built on the back of a truck bed by Renewable Oil International, shown in Figure 8.6. Other promising technologies for biofuel production, such as cellulosic ethanol production or alkane production via gasification and the Fischer–Tropsch Synthesis,

Fig. 8.6. A small-scale portable reactor for liquid fuel production by distributed fast pyrolysis (courtesy of Renewable Oil International).

are only economical at a large scale with feeds from 1000 to 8000 tons of biomass per day (Aden *et al.*, 2002; Hamelinck and Faaij, 2002; Hamelinck *et al.*, 2004; Hamelinck *et al.*, 2005; Prins *et al.*, 2004). Many researchers worldwide are studying fast pyrolysis of biomass, as shown in Table 8.1. A review of fast pyrolysis reactors has been given by Bridgwater (1999). Of the 24 reactors listed in Table 8.1, only 6 of them are in the United States, demonstrating that the United States has not yet put significant resources into studying fast pyrolysis. Many of the pyrolysis reactors listed in the table are being operated commercially.

One of the current challenges of fast pyrolysis is developing methods for conversion of bio-oils into liquid transportation fuels. Bio-oils cannot be used in conventional gasoline and diesel engines without upgrading. The bio-oils are chemically unstable and acidic (pH ~2.5). The acidity causes problems in diesel and gasoline engines. Pyrolysis-derived oils have a higher oxygen content, a higher moisture content, and a lower heating value (17 MJ/kg) than conventional fuel oil (43 MJ/kg). There are

Table 8.1. Fast pyrolysis reactor types and locations worldwide (adapted from Mohan et al., 2006).

Reactor Types	Locations
Ablative	Aston University (UK); NREL (US); BBC (Canada); Castle Capital (Canada)
Auger	Renewable Oil International (US); Mississippi State (US); Advanced Biorefinery, Inc. (Canada)
Circulating fluidized bed	CPERI (Greece), CRES (Greece), ENEL (Italy)
Entrained flow	GTRI (US), Egemin (Belgium)
Fluidized bed	Aston University (UK), Dynamotive (Canada); Hamburg University (Germany); Leeds University (UK); NREL (US); Oldenburg University (Denmark); VTT (Finland)
Rotating cone	University of Twente (Netherlands); BTG/Schelde/Kara (Netherlands)

two methods that are being developed to improve bio-oils: (1) improve the fast pyrolysis process by catalyst addition; and (2) catalytic conversion of bio-oils into gasoline and diesel fuel.

KiOR (http://www.kior.com/, a privately funded joint venture) is developing a fast pyrolysis process that produces an improved bio-oil by the addition of catalysts to the process. In this process, catalytically accessible biomass is converted into a bio-crude suitable for further processing in new or existing oil refineries to gasoline and/or diesel fuel. Because of the enhanced accessibility of the biomass to the catalyst, the catalyst in BCC not only improves the secondary tar cracking, as in classical catalytic pyrolysis, but the catalyst also enhances the kinetics and selectivity of the primary conversion of the solid biomass. This allows for conversion at lower temperatures (improved energy efficiency), while producing improved-quality products. BCC can build on the long history of the technology of fluid catalytic cracking (FCC), the low cost workhorse of the oil refining industry.

Huber and coworkers at the University of Massachusetts-Amherst have shown that gasoline range aromatics can be produced from solid biomass-derived feedstocks when zeolite-based catalysts are introduced into the pyrolysis process (Carlson et al., 2008). This process involves rapidly

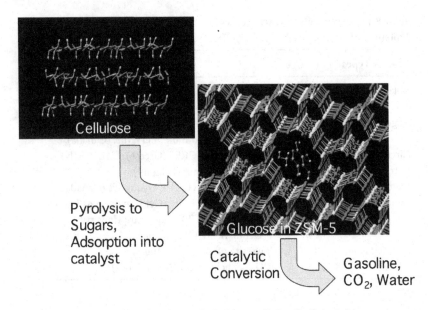

Fig. 8.7. Green gasoline by catalytic fast pyrolysis of cellulosic biomass.

heating the biomass, where pyrolysis of the biomass to smaller organic fragments occurs (Figure 8.7). These molecules then enter the zeolite catalysts, where they are converted into aromatics, CO, water, and coke. The aromatic yield is highly dependent on both the active site and the pore structure of the catalyst.

One option for conversion of pyrolysis oils into liquid fuels is to use existing petroleum refinery technology, including hydrotreating and catalytic cracking (Huber and Corma, 2007; Marinangeli et al., 2006). Some of the advantages of using existing infrastructure is that this results in lower capital costs. WTEC panelists who visited the Institute of Research on Catalysis and the Environment in Lyon, France (IRCELYON), heard about its project on hydrotreating and catalytic cracking of bio-oils, where its researchers are studying both model and real compounds. A recent review by Elliott (2007) discusses the historical developments of hydrotreating of bio-oils. At the time of this writing, bio-oils are not being used in a commercial-scale petroleum refinery. However, UOP (http://www.uop.com/) and Pacific Northwest National Laboratory (PNNL; http://www.pnl.gov/) have analyzed the opportunities of using bio-oils in existing petroleum refineries (Marinangeli et al., 2006). NESTE Oil and Shell are also studying how they could process bio-oils in petroleum refineries.

8.3.5 Liquid-phase/aqueous-phase catalytic processing

The high functionality and low volatility of biomass make it particularly amenable to liquid-phase (or aqueous-phase) processing; this method is particularly advantageous because thermally unstable compounds can be converted selectively to liquid fuels. A wide variety of products can be formed by liquid phase processing, including liquid alkanes (diesel fuel replacement), light alkanes (gasoline range alkanes), gaseous alkanes (natural gas), and hydrogen (Huber and Corma, 2007; Cortright et al., 2002; Huber et al., 2005; Huber and Dumesic, 2006; Huber et al., 2003; Huber et al., 2006). The overarching goal of liquid-phase catalytic processing of biomass-derived compounds (e.g., sugars) to produce liquid transportation fuels is to produce next-generation liquid fuels that (1) can be used with the existing infrastructure, (2) do not involve energetically intense distillation steps, and (3) have high rates of production per reactor volume. The next-generation liquid fuels produced by liquid-phase processing should have higher energy densities per volume (30–40 MJ/L) and have properties that are closer to those of gasoline and diesel fuel than ethanol (19 MJ/L).

High yields of well-defined liquid fuels are produced by liquid-phase catalytic processing. Liquid-phase catalytic processing of sugars is typically carried out at lower temperatures (e.g., 500 K) compared to biomass pyrolysis, liquefaction, or gasification. However, whereas these latter processes can operate with complex biomass feedstocks (e.g., containing cellulose, hemi-cellulose, and lignin components), liquid-phase catalytic processing typically involves feedstocks containing specific biomass-derived compounds such as sugars or polyols. Thus, another advantage of liquid-phase catalytic processing is that high selectivities and yields to targeted fuel compounds can be achieved, whereas a disadvantage of such processing is that real biomass feedstocks must be pretreated to prepare a feed solution that is appropriate for subsequent liquid-phase catalytic processing. Aqueous-phase processing work is being pioneered by James Dumesic at the University of Wisconsin (Cortright *et al.*, 2002; Huber *et al.*, 2003; 2004; 2005; Huber and Dumesic, 2006; Chheda *et al.*, 2007; Davda *et al.*, 2005), and is being commercialized by Virent Energy Systems (http://www.virent.com/). A number of labs worldwide are starting to study aqueous-phase processing, including the University of Munich and Aristotle University of Thessaloniki (Greece) (Kechagiopoulos *et al.*, 2006), and Fudan University (PR China) (Xie *et al.*, 2006).

8.3.6 *Vegetable oil conversion*

The production of biodiesel by transesterification of vegetable oils is a technology that is experiencing almost exponential growth, as shown in Figure 8.8. Mineral bases are typically used as catalysts for the transesterification process (Huber *et al.*, 2006; Lopez *et al.*, 2005; Lotero *et al.*, 2005). These mineral bases introduce complex separation problems into the transesterification process. Heterogeneous catalysts are being developed to replace mineral bases in many labs worldwide. It has been projected that

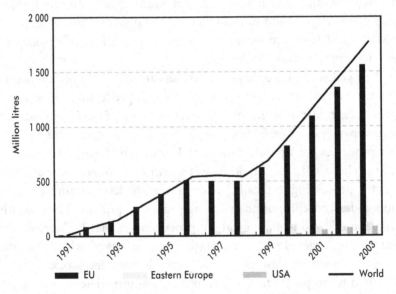

Fig. 8.8. World and regional biodiesel capacity from 1991–2003 (IEA, 2004).

the development of a solid base catalyst for the transesterification step could reduce the cost of biodiesel by 30 cents per gallon (Goodwin *et al.*, 2005). During their visits the WTEC panelists observed that there is a significant research effort in Europe on transesterification with solid base catalysts, including at IRCELYON and the Institut Charles Gerhardt Montpellier (ICGM) in France and at the Instituto de Tecnología Química (ITQ) in Spain. Most of the work at these institutions has not yet been published. In addition to the development of solid base catalysts for transesterification reactions, solid acid catalysts are also being developed for the esterification step in biodiesel synthesis (Lopez *et al.*, 2005; Lotero *et al.*, 2005; Mbaraka and Shanks, 2005). This is an important step for biodiesel production if the vegetable oils contain large amounts of free fatty acids. Researchers in the United States are also developing solid catalysts for biodiesel production.

Hydrotreating of vegetable oils to produce "green diesel" in existing petroleum refineries is another option for vegetable oil conversion that is moving to commercial scale (Huber and Corma, 2007; Marinangeli *et al.*, 2005; Huber *et al.*, 2007). This technology is currently using standard hydrotreating catalysts that have been developed for petroleum oil feedstocks; in the future it is likely that new catalysts will be developed that are specifically made for vegetable oil feedstocks. Neste Oil has developed a NExBTL[1] renewable diesel production technology based on hydrotreating and is building two new units onto its Porvoo refinery in Finland, each of which has an annual capacity of 170,000 tons. Neste Oil will also be building a similar unit at Total's (http://www.total.com/) refinery in the UK. ConocoPhillips (http://www.conocophillips.com) has processed waste vegetable oil with petroleum oil in a unit of its Humber Refinery in the UK. ConocoPhillips would like to also use this technology in the United States; however, the US government does not currently give the same subsidy for green diesel that it gives for biodiesel. UOP has also studied hydrotreating of vegetable oils in a petroleum refinery (Marinangeli *et al.*, 2006). This author knows of no academic labs in the United States that are studying the hydrotreating of vegetable oils.

---

[1] BTL = biomass to liquid.

8.3.7 Chemicals from biomass

Another area that is of growing importance is the production of chemicals from biomass. This area will become more important as more liquid fuels are produced from biomass, because byproducts from the fuel process can provide an inexpensive raw material that can be used to make chemicals. An example of this is the decreased cost and increased production of glycerol, which is a by-product of biodiesel refineries. Glycerol is being studied as a feedstock for production of a range of fuels and chemicals. A number of companies worldwide are starting to use bioplastics in their products. For example, Toyota uses polylactic acid in its Prius and Raum vehicles. Toyota is heavily investing in polylactic acid and hopes to control two-thirds of the polylactic acid market by 2020.

The conversion of biomass into chemicals involves the selective transformation of biomass-derived feedstocks. This involves a number of key fundamental reactions, including hydrogenation, dehydration, oxidation, hydrolysis, isomerization, and reforming (Chheda et al., 2007). These fundamental reactions can be combined with multifunctional catalysts where multiple reactions occur in a single reactor, such as in hydrogenolysis and dehydration/hydrogenation reactions.

The WTEC panel discussed the following biochemical projects with various hosts in Europe and Asia:

(1) Hydrogenation and selective oxidation of sugars at IRCELYON. Similar research is also being done in the United States at the University of Virginia.
(2) Furfural and 5-hydroxymethylfurfural (HMF) production by dehydration of sugars at Marseille. The group of Dumesic and researchers at PNNL are studying methods for furfural and HMF production (Chheda et al., 2007; Roman-Leshkov et al., 2006; Zhao et al., 2007).
(3) The production of sorbitol from cellulose (Hokkaido University Catalysis Research Center). Similar work is being done at University of Texas at San Antonio.
(4) Production of 1, 2 propanediol by hydrogenolysis of glycerol (University of Tsukuba). Members of the Kunimori-Tomishige group have studied the hydrogenolysis of glycerol to 1, 2 propanediol and 1, 3 propanediol. They have shown that this reaction occurs at lower

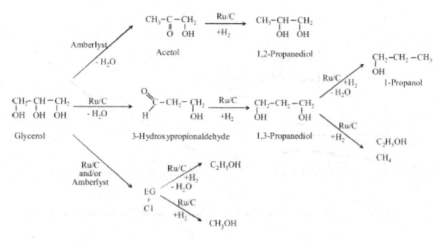

Fig. 8.9. Reaction scheme for conversion of glycerol over Ru/C and amberlyst-based catalysts (from Miyazawa et al., 2006).

temperatures with Ru/C + amberlyst (Miyazawa et al., 2006). They have also studied the reaction pathway (see Figure 8.9) and identified acetol as a key intermediate. They identified Rh/SiO$_2$ as a potential catalyst for 1, 3 propanediol production (Furikado et al., 2007). They are now starting to develop hydrogenolysis reactions for other biomass-derived oxygenated molecules. Similar work is also being done throughout the United States, including at Iowa State University, Michigan State University, the University of Missouri, and the University of Virginia.

(5) Glycerol conversion by catalytic cracking (ITQ, Spain) (Huber and Corma, 2007; Corma et al., 2007 and n.d.). Similar work is being done by Daniel Resasco at the University of Oklahoma.

8.3.8 Bibliometric analysis of catalysis and biofuels

WTEC commissioned a bibliometric analysis of catalysis papers that are related to biomass conversion. The full details of the bibliometric study are reported in Site Reports-Europe on the International Assessment of Research in Catalysis by Nanostructured Materials website, www.wtec.org/ private/catalysis (password required; contact WTEC). As shown in Figure 8.10, from 1996–2005 only 1 out of every 400 catalysis papers was

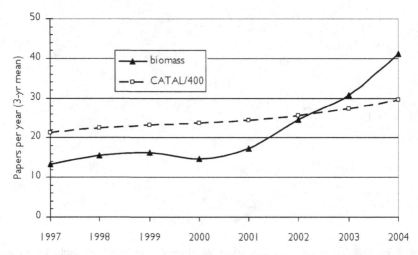

Fig. 8.10. Catalysis papers (divided by 400) compared with catalysis papers relevant to biomass as a source of energy (biomass), 1996–2005, three-year running means (see also Site Reports-Europe on the International Assessment of Research in Catalysis by Nanostructured Materials website, www.wtec.org/private/catalysis (password required; contact WTEC)).

related to biomass conversion. However, in 2002 the number of papers published in this area started growing rapidly. Based on the results of the WTEC international assessment, it would seem that this area is continuing to grow worldwide.

The WTEC bibliometric study also indicated that less than 12% of the papers in catalysis related to biofuels came from the United States for the same time period, as shown in Figure 8.11. The majority of these papers came from the combined countries of Europe. This indicates that the United States is not putting as many resources into biofuel-related catalysis projects as Europe and Japan.

8.4 Photocatalytic Water Splitting

Another application of catalysis to renewable energy is the photocatalytic splitting of water with visible light to hydrogen and oxygen. This research work began with Honda and Fujishima in 1970, who demonstrated water splitting with a photoelectrochemical (PEC) cell (with a TiO_2 anode and

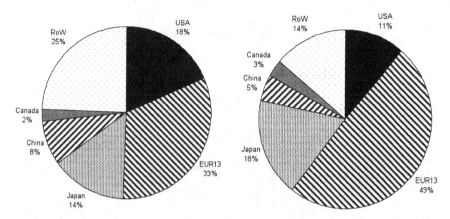

Fig. 8.11. Geographical distribution of all catalysis papers (*left*) and catalysis papers related to biomass conversion (*right*), for the years 1996–2005 (see Site Reports-Europe on the International Assessment of Research in Catalysis by Nanostructured Materials website, www.wtec.org/private/catalysis (password required; contact WTEC)).

Pt cathode) under UV irradiation and an external bias (Fujishima and Honda, 1972). Several researchers that the WTEC panelists visited during the course of this WTEC study are working on photocatalytic splitting of water, including Domen (University of Tokyo); Fierro (Institute of Catalysis and Petrochemistry in Madrid); Hutchings (Cardiff University, UK); and several groups at the National Institute of Advanced Industrial Science and Technology (AIST) in Japan.

The basic principle of overall water splitting on a heterogeneous catalyst is shown in Figure 8.12. The overall water splitting is a large positive change in Gibbs free energy. Electrons in the valence band are excited into the conduction band by irradiation at an energy equivalent to or greater than the band gap of the photocatalyst, producing holes in the valence band. These electron-hole pairs cause reduction and oxidation reactions, respectively.

Domen and coworkers have developed a heterogeneous photocatalyst that can produce hydrogen from visible light in a single step, as shown in Figure 8.13 (Maeda and Domen, 2007). The challenges associated with developing a suitable photocatalyst include finding materials that have the following properties: (1) band edge potentials suitable for overall water splitting; (2) band gap energy smaller than 3 eV; and (3) stability in the photocatalytic reaction.

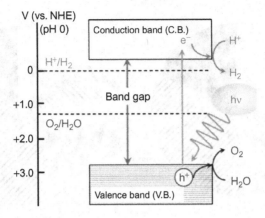

Fig. 8.12. Basic principles of overall water splitting on a heterogeneous photocatalyst (from Maeda and Domen, 2007).

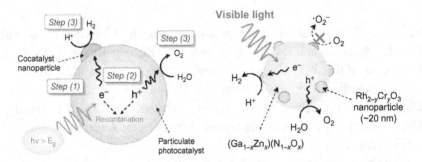

Fig. 8.13. Heterogeneous photocatalyst for water splitting developed by Domen *et al.*, (from Maeda and Domen, 2007).

8.5 Conclusions

Using heterogeneous catalysis to convert renewable resources into liquid fuels, chemicals, and energy is currently a small field that appears to be growing exponentially. As fossil fuel prices continue to increase and as we look to solve problems caused by man-made CO_2 emissions, heterogeneous catalysis is likely to be part of the answer. There is a broad range of catalysis projects related to conversion of renewable resources, including different types of biomass and sunlight, into fuels and chemicals. It is likely that in the future catalysis will be used to convert our renewable resources into fuels and chemicals, just as today catalysis converts our nonrenewable resources into fuels and chemicals.

References

Aden, A., Ruth, M., Ibsen, K., Jechura, J., Neeves, K., Sheehan, J., Wallace, B., Montague, L., Slayton, A., and Lukas, J. (2002). *Lignocellulosic Biomass to Ethanol Process Design and Economics Utilizing Co-Current Dilute Acid Prehydrolysis and Enzymatic Hydrolysis for Corn Stover* (National Renewable Energy Laboratory, Golden, CO), Report NREL/TP-510-32438.

Asadullah, M., Ito, S., Kunimori, K., Yamada, M., and Tomishige, K. (2002). Biomass gasification to hydrogen and syngas at low temperature: Novel catalytic system using fluidized bed reactor, *J. Catal.*, 208, pp. 255–259.

Boerrigter, H. (2003). "Green" Diesel Production with Fischer–Tropsch Synthesis (Energy Research Centre of the Netherlands, Petten, Holland), Report #ECN-RX—03–014, http://www.ecn.nl/publications/.

Boerrigter, H., den Uil, H., and Calis, H.-P. (2002). Green diesel from biomass via Fischer–Tropsch synthesis: New insights in gas cleaning and process design, *Proc. Pyrolysis and Gasification of Biomass and Waste Expert Meeting*.

Bridgwater, A. V. (1999). Principles and practice of biomass fast pyrolysis processes for liquids, *J. Anal. App. Pyrol.*, 51, pp. 3–22.

Bridgwater, A. V., and Peacocke, G. V. C. (2000). Fast pyrolysis processes for biomass, *Renew. Sust. Energ. Rev.*, pp. 41–73.

Carlson, T., Vispute, T., and Huber, G. W. (2008). Green gasoline by catalytic fast pyrolysis of solid biomass-derived compounds, *Chem. Sus. Chem.*, 1(5), p. 369.

Chheda, J. N., Huber, G.W., and Dumesic, J. A. (2007). Liquid-phase catalytic processing of biomass-derived oxygenated hydrocarbons to fuels and chemicals, *Ange. Chem. Int. Ed.*, 46(38), pp. 7164–7183.

Chheda, J. N., Roman-Leshkov, Y., and Dumesic, J. A. (2007). Production of 5-hydroxymethylfurfural and furfural by dehydration of biomass-derived mono- and poly-saccharides, *Green Chem.*, 9, pp. 342–350.

Corma, A., Huber, G. W., Sauvanaud, L., and O'Connor, P. (2007). Processing biomass-derived oxygenates in the oil refinery: Catalytic cracking (FCC) reaction pathways and role of catalyst, *J. Catal.*, 247, pp. 307–327.

Corma, A., Huber, G. W., Sauvanaud, L., and O'Connor, P. (n.d.) Production of oxygenates from catalytic dehydration of glycerol in a fluid catalytic cracking process. In preparation.

Cortright, R. D., Davda, R. R., and Dumesic, J. A. (2002). Hydrogen from catalytic reforming of biomass-derived hydrocarbons in liquid water, *Nature*, 418, pp. 964–967.

Davda, R. R., Shabaker, J. W., Huber, G. W., Cortright, R. D., and Dumesic, J. A. (2005). A review of catalytic issues and process conditions for renewable hydrogen and alkanes by aqueous-phase reforming of oxygenated hydrocarbons over supported metal catalysts, *App. Cat. B*, 56, pp. 171–186.

de la Torre Ugarte, D. G., Walsh, M. E., Shapouri, H., and Slinsky, S. P. (2003). *The Economic Impacts of Bioenergy Crop Production on US Agriculture* (USDA, Washington, DC), http://bioenergy.ornl.gov/papers/misc/eco_impacts.html.

Elliott, D. C. (2007). Historical developments in hydroprocessing bio-oils, *Energy and Fuels*, 21, pp. 1792–1815.

Energy Information Administration, US Department of Energy (2008). *Annual Energy Outlook* (Department of Energy, Washington, DC), http://www.eia.doe.gov/oiaf/aeo/.

Fujishima, A., and Honda, K. (1972). Electrochemical photolysis of water at a semiconductor electrode. *Nature*, 238, pp. 37–38.

Furikado, I., Miyazawa, T., Koso, S., Shimao, A., Kunimori, K., and Tomishige, K. (2007). Catalytic performance of Rh/SiO$_2$ in glycerol reaction under hydrogen, *Green Chem.*, 9, pp. 582–588.

Goodwin, J. G., Bruce, D. A., Lotero, E., Mo, X., Liu, Y., Lopez, D. E., Suwannakarn, K., Gangwal, S., McMichael, W. J., and Green, D. (2005). Heterogeneous catalyst development for biodiesel synthesis, *Ind. Eng. Chem. Res.*, 44(14), pp. 5353–5363.

Hamelinck, C. N., and Faaij, A. P. C. (2002). Future prospects for production of methanol and hydrogen from biomass, *J. Power Sources*, 111, pp. 1–22.

Hamelinck, C. N., Faaij, A. P. C., den Uil, H., and Boerrigter, H. (2004). Production of FT transportation fuels from biomass; Technical options, process analysis and optimisation, and development potential, *Energy*, 29, pp. 1743–1771.

Hamelinck, C. N., van Hooijdonk, G., and Faaij, A. P. (2005). Ethanol from lignocellulosic biomass: Techno-economic performance in short-, middle- and long-term, *Biomass and Bioenergy*, 28, pp. 384–410.

Huber, G. W., Chheda, J. N., Barrett, C. J., and Dumesic, J. A. (2005). Production of liquid alkanes by aqueous-phase processing of biomass-derived carbohydrates, *Science*, 308, pp. 1446–1450.

Huber, G. W., and Corma, A. (2007). Synergies between bio- and oil refineries for the production of fuels from biomass, *Angew. Chem. Int. Ed. Engl.*, 46(38), pp. 7184–7201.

Huber, G. W., Cortright, R. D., and Dumesic, J. A. (2004). Renewable alkanes by aqueous-phase reforming of biomass-derived oxygenates, *Angew. Chem.*, 43, pp. 1549–1551.

Huber, G. W., and Dumesic, J. A. (2006). An overview of aqueous-phase catalytic processes for production of hydrogen and alkanes in a biorefinery, *Catal. Today*, 111(1–2), pp. 119–132.

Huber, G. W., Iborra, S., and Corma, A. (2006). Synthesis of transportation fuels from biomass: Chemistry, catalysts and engineering, *Chem. Rev.*, 106, pp. 4044–4098.

Huber, G. W., O'Connor, P., and Corma, A. (2007). Processing biomass in conventional oil refineries: Production of high quality diesel by hydrotreating vegetable oils in heavy vacuum oil mixtures, *App. Catal. A*, 329, pp. 120–129.

Huber, G. W., Shabaker, J. W., and Dumesic, J. A. (2003). Raney Ni-Sn catalyst for H_2 production from biomass-derived hydrocarbons, *Science*, 300, pp. 2075–2077.

Huber, G. W., Shabaker, J. W., Evans, S. T., and Dumesic, J. A. (2006). High throughput studies of supported Pt and Pd bimetallic catalysts for aqueous phase reforming of ethylene glycol, *App. Cat. B*, 62, pp. 226–235.

International Energy Agency (2004). *Biofuels for Transport: An International Perspective* (International Energy Agency, Paris, France), http://www.iea.org/Textbase/publications/index.asp.

Kechagiopoulos, P. N., Voutetakis, S. S., Lemonidou, A. A., and Vasalos, I. A. (2006). Hydrogen production via steam reforming of the aqueous phase of bio-oil in a fixed bed reactor, *Energy and Fuels*, 20, pp. 2155–2163.

Klass, D. L. (1998). *Biomass for Renewable Energy, Fuels, and Chemicals* (Academic Press, San Diego).

Liu, Y. L. (2007). Chinese biofuels expansion threatens ecological balance, *Ren. Energy Access*, http://www.worldwatch.org/node/4959.

Lopez, D. E., Goodwin, J. G., Jr., Bruce, D. A., and Lotero, E. (2005). Transesterification of triacetin with methanol on solid acid and base catalysts, *Appl. Catal. A*, 295, pp. 97–105.

Lotero, E. Liu, Y., Lopez, D. E., Suwannakarn, K., Bruce, D. A., and Goodwin, J. G. (2005). Synthesis of biodiesel via acid catalysis, *Ind. Eng. Chem. Res.*, 44, pp. 5353–6363.

Maeda, K., and Domen, K. (2007). New non-oxide photocatalysts designed for overall water splitting under visible light, *J. Phys. Chem, C*, 111, pp. 7851–7861.

Marinangeli, R., Marker, T., Petri, J., Kalnes, T., McCall, M., Mackowiak, D., Jerosky, B., Reagan, B., Nemeth, L., Krawczyk, M., Czernik, S., Elliott, D., and Shonnard, D. (2006). *Opportunities for Biorenewables in Oil Refineries* (UOP, Des Plaines, IL), DOE Report No. DE-FG36-05GO15085.

Mbaraka, I. K., and Shanks, B. H. (2005). Design of multifunctionalized mesoporous silicas for esterification of fatty acid, *J. Catal.*, 229, pp. 365–373.

Miyazawa, T., Kusunoki, Y., Kunimori, K., and Tomishige, K. (2006). Glycerol conversion in the aqueous solution under hydrogen over Ru/C + an ion-exchange resin and its reaction mechanism, *J. Catal.*, 240, pp. 213–221.

Mohan, D., Pittman, C. U., and Steele, P. H. (2006). Pyrolysis of wood/biomass for bio-oil: A critical review, *Energy and Fuels*, 20, pp. 848–889.

[No Author] (2007). Neste Oil and Stora Enso to partner on biomass-to-liquids from wood residues, *Renewable Energy Access*.

Perlack, R. D., Wright, L. L., Turhollow, A., Graham, R. L., Stokes, B., and Erbach, D. C. (2005). Biomass as feedstock for a bioenergy and bioproducts industry: The technical feasibility of a billion-ton annual supply, US Department of Energy and U.S. Department of Agriculture Forest Service, Washington, DC.

Prins, M. J., Ptasinski, K. J., and Janssen, F. J. J. G. (2004). Exergetic optimization of a production process of Fischer–Tropsch fuels from biomass, *Fuel Proces. Tech.*, 86, pp. 375–389.

Roman-Leshkov, Y., Chheda, J. N., and Dumesic, J. A. (2006). Phase modifiers promote efficient production of hydroxymethylfurfural from fructose, *Science*, 312, pp. 1933–1937.

Salge, J. R., Dreyer, B. J., Dauenhauer, P. J., and Schmidt, L. D. (2006). Renewable hydrogen from nonvolatile fuels by reactive flash volatilization, *Science*, 314, pp. 801–804.

Xie, F., Chu, X., Hu, H., Qiao, M., Yan, S., Zhu, Y., He, H., Fan, K., Li, H., Zong, B., and Zhang, X. (2006). Characterization and catalytic properties of Sn-modified rapidly quenched skeletal Ni catalysts in aqueous-phase reforming of ethylene glycol, *J. Catal.*, 241, pp. 211–220.

Zhao, H., Holladay, J. E., Brown, H., and Zhang, Z. C. (2007). Metal chlorides in ionic liquid solvents convert sugars to 5-hydroxymethylfurfural, *Science*, 316, pp. 1597–1600.

Appendix 1: Panelists' Biographies

A.1 Robert J. Davis, PhD; Panel Chair

PhD, Stanford University (Chemical Engineering, 1989)
MS, Stanford University (Chemical Engineering, 1987)
BS, Virginia Tech (Chemical Engineering, 1985)

Robert Davis has been professor and chair of Chemical Engineering (CHE Department) at the University of Virginia since 2002. After obtaining his PhD in 1989, he worked as a postdoctoral research fellow in the Chemistry Department at the University of Namur in Belgium. He joined the faculty in Chemical Engineering at the University of Virginia as an assistant professor in 1990, was promoted to associate professor in 1996, and to full professor in 2002.

Professor Davis has extensively used *in situ* spectroscopic methods coupled with both steady-state and transient kinetic methods to elucidate how oxide supports and basic promoters alter the active sites for a variety of catalytic reactions, including selective oxidation of hydrocarbons, acid/base reactions, and ammonia synthesis. Most recently, he has focused on the catalytic conversion of biorenewable carbon sources to fuels and chemicals. He and his group have illustrated how changes in the elemental composition, atomic structure, and reaction environment in the vicinity of a catalytic site influence both the activity and selectivity of chemical reactions.

He received the 2007 Paul H. Emmett Award in Fundamental Catalysis from the North American Catalysis Society, the NSF Young Investigator Award, the DuPont Young Professor Award, the Union Carbide Innovation Recognition Award, and the UVa Rodman Scholars Award for Excellence in Teaching. He has co-authored about 100 publications, 1 patent, and 1 textbook, *Fundamentals of Chemical Reaction Engineering*. He has delivered 100 invited lectures at conferences, academic departments, and industrial research groups, and has co-authored over 90 additional presentations at technical meetings.

Professor Davis has served as President of the Southeastern Catalysis Society, Chair of the 2006 Gordon Research Conference on Catalysis, Chair of Catalysis Programming of the AIChE, Director of the Catalysis and Reaction Engineering Division of the AIChE, Co-Chair of an International Catalysis Workshop in China, member of the Advisory Board of the International Conferences on Solid Acid and Base Catalysis, and member of the editorial boards of *Applied Catalysis A and B* and the *Journal of Molecular Catalysis A*.

A.2 Vadim V. Guliants, PhD

PhD, Princeton University (Chemistry, 1995)
MA, Princeton University (Chemistry, 1993)
Undergraduate, Moscow State University (Chemistry, 1987)

Vadim Guliants is professor in the Department of Chemical and Materials Engineering (CME) at the University of Cincinnati. He obtained his diploma in Chemistry (specializing in Physical Chemistry) with highest honors from Moscow State University (USSR) in 1987. Subsequently, he continued his studies in the Departments of Chemistry and Chemical Engineering at Princeton University. His PhD research centered on synthesis and characterization of novel vanadium-phosphorus-oxides for the partial oxidation of n-butane. Subsequent to receiving his PhD from Princeton, he joined R&D at Praxair, Inc., where his research focused on molecular modeling, synthesis, and characterization of microporous frameworks for gas separation. Vadim Guliants moved to the University of Cincinnati (UC) in 1999 as Assistant Professor of Chemical Engineering, where he was promoted to associate and full professor in 2003 and 2005, respectively. Professor Guliants also served as Associate Department Head of the Department of Chemical and Materials Engineering in 2004–2005.

Professor Guliants's research directions are in the areas of (1) mixed metal Mo-V-based oxides for partial (amm)oxidation of lower alkanes, (2) novel nanostructured Ni based catalysts for partial oxidation and reforming of renewable and fossil feedstocks, and (3) nanostructured biocatalytic membranes employing entrapped lipases and monoxygenases for chemical processing and other applications. The unifying theme of his research is elucidating the fundamental relationships between the molecular structure of these materials and their function for the purpose of rational design of new improved functional materials for applications spanning chemical and environmental catalysis and separations.

He has received the NSF CAREER Award, NSF NIRT Award, University of Cincinnati College of Engineering Outstanding Research Award for Young Faculty, University of Cincinnati Teaching Diversity Award, and a Soros Scholarship at Oxford University. He has co-authored about 80 publications and 6 patents. He has delivered over 50 invited lectures at conferences, academic departments, and industrial research groups, and has co-authored over 90 additional presentations at technical meetings.

Professor Guliants has served as Director of the TriState (OH, KY, WV) Catalysis Society, Chair of Ceramic Materials Programming of the AIChE, Secretary of the ACS Catalysis Secretariate, and a guest editor of several issues of *Catalysis Today* and *Topics in Catalysis*.

A.3 George W. Huber, PhD

PhD, University of Wisconsin, Madison (Chemical Engineering, 2005)
MS, Brigham Young University (2000)
BS, Brigham Young University (1999)

George Huber is an assistant professor (and Armstrong Professor) of Chemical Engineering at the University of Massachusetts-Amherst. His research focus is on breaking the chemical and engineering barriers to lignocellulosic biofuels. Prior to his appointment at UMass, Dr. Huber did a post-doctoral stay with Avelino Corma at the Technical Chemical Institute at the Polytechnical University of Valencia, Spain (UPV-CSIC), where he studied biofuels production using petroleum-refining technologies. While obtaining his PhD in Chemical Engineering from University of Wisconsin-Madison, he helped develop novel aqueous-phase catalytic processes for biofuels production under the guidance of Jim Dumesic. At Brigham Young University where he obtained his BS and MS degrees and studied the Fischer–Tropsch Synthesis, Calvin H. Bartholomew was his MS advisor.

Dr. Huber has authored 25 peer-reviewed publications, including two papers in *Science* and three articles in *Angewandte Chemie International Edition*. He has five patents in the area of biofuels. His research is being commercialized by two different companies (Virent Energy Systems and Bio-e-con). He is currently working with governmental and industrial institutions to help make cellulosic biofuels a reality. He is the chair for an NSF/DOE workshop entitled, "Breaking the Chemical and Engineering Barriers to Lignocellulosic Biofuels" (http://www.ecs.umass.edu/biofuels).

A.4 Raul F. Lobo, PhD

PhD, California Institute of Technology (Chemical Engineering, 1995)
MSc, California Institute of Technology (Chemical Engineering, 1993)
Licenciatura, University of Costa Rica (Chemical Engineering, 1989)

Raul Lobo is full professor in the Department of Chemical Engineering and is Associate Director of the Center for Catalytic Science and Technology at the University of Delaware. His research interests span the development of novel porous materials for catalysis and separations, the characterization of disordered zeolite structures, and the scientific aspects of catalysts synthesis. After receiving his PhD in the study of new molecular sieves and zeolites from CalTech, he worked for one year at Los Alamos National Laboratory, New Mexico, as a postdoctoral fellow. He started his academic career at the Center for Catalytic Science and Technology, University of Delaware, in the fall of 1995.

Professor Lobo has published over seventy refereed reports and he is co-inventor in three US patents. He has been recipient of the NSF CAREER award, the Camille Dreyfus Teacher-Scholar award, the Young Scholar Award of the Francis Alison Society, the Outstanding Young

Faculty Award of the College of Engineering (University of Delaware), the Innovation Recognition Program Award of the Dow Chemical Company, and the Ipatieff Prize of the American Chemical Society.

Professor Lobo has been a member of the Structure Commission of the International Zeolite Association since 1999, and he is a member of editorial board of *TIP Revista Especializada en Ciencias Químico-Biológicas, México*. He has been a visiting fellow at the Centre for High-Resolution Electron Microscopy at Delft University of Technology in The Netherlands. In 1997 he was a member of the executive committee of the International Zeolite Conference, and in 2003 he organized the Symposium Transmission Electron Microscopy and Catalysis and was chair of the Catalysis and Reaction Engineering Division of AIChE.

A.5 Jeffrey T. Miller, PhD

PhD, Oregon State University (Chemistry, 1980)
MS, University of New Mexico (Inorganic Chemistry, 1973)
BS, Memphis State University (Chemistry, 1971)

Jeff Miller is Senior Scientist in the heterogeneous catalysis group at the Argonne National Laboratory, developing testing and characterization of new catalytic materials for energy production. His focus is on characterization of catalysts under realistic reaction conditions of high temperature and pressure, especially via use of EXAFS spectroscopy at the Advanced Photon Source. He is also an adjunct professor in the Department of Chemical Engineering and a member of the College of Engineering Advisory Board of the University of Illinois at Chicago; in addition, he has been a mentor at Northwestern and Purdue universities.

Prior to the summer of 2008, Dr. Miller was Senior Research Scientist at BP Chemical Company. After earning his PhD in 1980 he joined the exploratory catalyst and process development group of Amoco Oil Company (now BP) in Naperville, IL. His work at BP Chemical Company was focused on developing new catalytic materials for new chemical plants. His research interests at BP included synthesis, characterization, and applications of zeolite and nanoparticle heterogeneous catalysts, especially the use of *in situ* X-ray absorption spectroscopy for elucidation of the role of the support, size, and composition on the changes in structure and electronic properties of metal nanoparticles, and how these alter performance.

Dr. Miller received the 2006 Herman Pines Award for outstanding research in the field of catalysis from the Chicago Section of the North American Catalysis Society, for which he has also served as area director. Through collaborations with several leading academic groups in the United States and Europe, Dr. Miller has co-authored over 100 publications and been granted 45 US patents. Additionally, he has been on the editorial board of *Applied Catalysis A: General* and has been on the organizational committees for several international meetings.

A.6 Matthew Neurock, PhD

PhD, University of Delaware (Chemical Engineering, 1992)
BS, Michigan State University (Chemical Engineering, 1986)

Matt Neurock is Alice M. and Guy A. Wilson Professor of Chemical Engineering and professor of Chemistry at the University of Virginia. From 1992–1993, he was a postdoctoral fellow at the Eindhoven University of Technology in The Netherlands. He returned to the United States in 1993 to work for the DuPont Chemical Company as a visiting scientist in its Corporate Catalysis Center. He joined the faculty of Chemical Engineering at the University of Virginia in 1995 as an assistant professor and was promoted to full professor in Chemical Engineering and in Chemistry in 2003. His research interests include computational heterogeneous catalysis, molecular reaction engineering of complex systems, surface bonding, and reactivity. He has developed and applied *ab initio* as well as stochastic methods to simulate the kinetic behavior of complex catalytic systems, including electrocatalysis, oxygenate synthesis, alkane activation, exhaust emissions, solid acid catalysis, the conversion of biorenewables, and catalyst deactivation.

Professor Neurock received the 2005 Paul H. Emmett Award in Fundamental Catalysis from the North American Catalysis Society. He was named the Johansen–Crosby Lecturer by Michigan State University in 2006, Distinguished Catalysis Researcher by Pacific Northwest Laboratory in 2004, and Visiting Professor by the University of Montpellier in 2007. He has also been the recipient of an NSF Career Development Award, a DuPont Young Faculty Award, and a Ford Young Faculty Award. He has co-authored over 145 publications, two patents, and two books. His most recent book, published in March 2006, is titled, *Molecular Heterogeneous Catalysis: A Conceptual and Computational Approach.* He was recently appointed as an editor for the *Journal of Catalysis*.

A.7 Renu Sharma, PhD

PhD, University of Stockholm (Solid State Chemistry, 1985)
MS, University of Stockholm (Solid State Chemistry, 1984)
BS and BEd, Punjab University (Physics and Chemistry, respectively, 1970)

Renu Sharma is Director of the Industrial Associates Program and Research Scientist at the LeRoy Eyring Center for Solid State Science at Arizona State University (ASU). Subsequent to receiving her PhD in Sweden in 1985, she came to ASU as a faculty research associate. She was appointed as assistant research scientist in the Center for Solid State Science in 1989, was promoted to associate research scientist with tenure in 1995, and to senior research scientist in 2008. In 1992 she joined as affiliated faculty the Science and Engineering of Materials program, which is now the School of Materials. Her research is primarily focused on atomic-scale *in situ* observations of the nanoscale synthesis processes and effects of ambient conditions on the functioning of nanomaterials, such as catalyst nanoparticles, carbon nanotubes, and silicon nanowires. This is achieved using an environmental scanning transmission electron microscope (ESTEM), established under her leadership, that combines atomic-level observations with chemical analysis of reaction processes. She has successfully developed and employed this unique technique to obtain atomic-level mechanisms involving the reactivity of nanoparticles, including surfaces, defects, commensurate and non-commensurate structures in nonstoichiometric inorganic solids, and ceramics.

Dr. Sharma has received fellowships from Swedish Institute (1980), University of Stockholm (1981), and Deutscher Akademischer

Austauschdienst (DAAD) (2000). Her students have won best poster awards and travel grants from AIMS, MSA, and NAM in 1991, 2006 and 2007, respectively. She has edited one book and has published one book chapter and 60 articles. She has presented over 40 invited papers at various institutes and conferences and has co-authored over seventy presentations at various national and international conferences.

She is past president of the Arizona Imaging and Microanalysis Society. She has also organized a symposium at Materials Research Society (MRS) and chaired sessions at Microscopy Society of America and MRS meetings. She has participated in three workshops, organized by DOE (2001), NSF (2006) and NIST (2006), and has chaired one NSF-sponsored workshop (2006), all in the general areas of *in situ* electron microscopy and nanotechnology. She has served as guest editor of *Microscopy and Microanalysis*.

A.8 Levi T. Thompson, PhD

PhD, University of Michigan (Chemical Engineering, 1986)
MSE, University of Michigan (Chemical Engineering and Nuclear Engineering, 1983)
BChE, University of Delaware (Chemical Engineering, 1982)

Levi Thompson is Richard E. Balzhiser Collegiate Professor of Chemical Engineering, Professor of Mechanical Engineering, and Director of the Center for Catalysis and Surface Science at the University of Michigan, Ann Arbor. He is also Director of the Michigan-Louis Stokes Alliance for Minority Participation, a US$5 million, NSF-funded program that teams the University of Michigan, Western Michigan University, Michigan State University, and Wayne State University in an effort to significantly increase the number of minority students earning science, technology, engineering, and mathematics baccalaureate degrees. After earning his PhD and working for two years at KMS Fusion, Inc., he joined the faculty of the U. Michigan Department of Chemical Engineering in 1988 and was promoted to full professor in 2000. From 2001 to 2005, Professor Thompson served as Associate Dean for Undergraduate Education, in which capacity he led the College of Engineering's student recruitment and support efforts to promote excellence and diversity in the undergraduate student body.

Research in Prof. Thompson's group at the University of Michigan focuses primarily on the design and development of high-performance catalytic, electrocatalytic, and adsorbent materials, and on defining relationships between the structure, composition, and function of nanostructured catalytic and electrocatalytic materials. In addition, his group works on the use of micromachining and self-assembly methods to fabricate hydrogen production systems and fuel cells. Professor Thompson's publications include an invited contribution on batteries for the *World Book Encyclopedia* and review articles on carbide and nitride catalysts, and hydrogen production. He has been awarded 10 patents. In 2007, he was appointed founding director of the Hydrogen Energy Technology Laboratory, a unit of the Michigan Memorial Phoenix Energy Institute. This laboratory was established to support multidisciplinary hydrogen production, storage, and conversion research, and is housed in the Phoenix Memorial Laboratory.

Professor Thompson's honors and awards include the NSF Presidential Young Investigator Award, Engineering Society of Detroit Gold Award, Union Carbide Innovation Recognition Award, Dow Chemical Good Teaching Award, College of Engineering Service Excellence Award, and Harold Johnson Diversity Award. He is cofounder, with his wife Maria, of T/J Technologies, a developer of nanomaterials for advanced batteries, and in 2006 he founded Inmatech, Co., Ltd., to commercialize catalytic materials and processes discovered and developed in his Univ. of Michigan laboratories. Professor Thompson is Consulting Editor for the *AIChE Journal*, and he is a member of the External Advisory Committee for the Center of Advanced Materials for Purification of Water with Systems (NSF Science and Technology Center at the University of Illinois), the National Academy's Committee on the Assembled Chemical Weapons Alternatives Programs, the AIChE Chemical Engineering Technology Operating Council, and the National Academy's Chemical Sciences Roundtable.

Appendix 2:
Bibliometric Analysis of Catalysis Research, 1996–2005

Grant Lewison

B.1 Summary

This study examined the outputs of catalysis research recorded in the Science Citation Index. The USA was the largest producer, but its quantitative dominance of catalysis was much less than in many other scientific fields, and Western Europe published almost twice as much. The output of the People's Republic of China is expanding rapidly and is now greater than that of Japan. The potential citation impact of U.S. papers was the highest in the world, although that of some European countries, notably Switzerland, was almost as high. By contrast, East Asian papers tended to appear in low-impact journals. An analysis was also made of the small number of papers in biomass research involving catalysis. Leading catalysis research institutions in Europe and East Asia were identified as an aide to the panel's selection of sites to visit.

B.2 Introduction

There were two main objectives of this study: to compare US research outputs in the field of catalysis with those of other leading countries, both quantitatively and in terms of likely impact; and to help the panel identify appropriate laboratories for their visits and leading researchers within them.

Bibliometrics is the quantitative study of publications, more specifically of research papers in the serial literature, usually restricted to articles and reviews. It has been practiced for more than 80 years, but it was given great impetus in the early 1960s with the creation by Eugene Garfield of the Science Citation Index (SCI) and more recently by the advent of powerful computers that can analyze the bibliographic details of large numbers of papers. Because bibliometric analysis is a purely mechanical process, whereby individual scientific papers are categorized in various ways, it should not be seen as the only approach to research evaluation. Rather, it is a complement to peer review, and its conclusions need to be validated by those with experience of the subject under study. Nevertheless, its objectivity can often be advantageous, as it is hard for even dispassionate observers to be able to sum up the achievements of a whole country, and to put them in context.

This study examined world outputs in a single field, and so the first step was to define this field and then create a "filter" that would selectively identify relevant papers within a bibliographic database and allow their details to be downloaded to a spreadsheet. From the addresses on the papers, a geographical analysis could be conducted, with the outputs of different countries, and of cities and institutions within them, determined. There is an assumption often made in bibliometrics to the effect that the impact of a paper on other researchers will be correlated with the frequency with which papers in the same journal are cited by other papers. This is not true at the individual paper level, but the correlation between potential and actual citation impact is surprisingly high for large groups of papers, such as those from a particular country. Bibliometric analysis is quite good at ranking countries and institutions working within a field on the basis of the journals in which they are published. This represents the considered opinions of an editorial board and several referees who will have read the paper carefully before accepting it for publication. However it is clearly not the same as the direct effect of the paper on other researchers, whose citation of it is sometimes treated as the only useful measure of research quality.

B.3 Methodology

As mentioned above, the first step in the analysis was to define the subject area, or field, for investigation. The definition was written by Professor Bob Davis, the panel chairman, and agreed with other panelists. It reads as follows:

> Heterogeneous catalysis involves the selective acceleration of a chemical reaction on an active site located on the surface of a solid material. Although catalysis involves molecular change, the nanostructure surrounding the active site is critical for activity and selectivity of the catalyst. The influence of nanostructure on catalysis includes but is not limited to the following: modification of the electronic structure by a quantum size effect, molecular traffic control and shape selectivity in nanometer-sized pores, optimal interactions of promoters and supports with active sites, cooperativity of active sites in multifunctional catalysts, and the liquid structure of the reaction medium that can provide alternate paths for catalytic reactions.

Based on this definition, a filter was created in close consultation with Prof. Davis that consisted of five parts:

(a) Names of specialist catalysis journals
(b) Title words that alone indicated that the paper was relevant to catalysis
(c) Title words indicative of "activity"
(d) Title words indicative of an element or other structure used for catalysis
(e) Title words showing that the paper was not, in fact, relevant to catalysis as defined

Papers were selected for analysis if they met criteria A or B, or C + D, but did not contain E. Tables B.1 to B.5 show the journals and title words in the five lists, A to E. In Tables B.2 to B.5, an asterisk denotes any character(s), or none.

Appendix 2

Table B.1. Set A: Specialist catalysis journals.

Advanced Catalysis	Journal of Catalysis
Applied Catalysis A General	Journal of Molecular Catalysis A Chemical
Applied Catalysis B Environmental	Journal of Molecular Catalysis B Enzymatic
Catalysis Reviews Science & Engineering	Kinetic Catalysis (English Translation)
Catalysis Surveys from Asia	Microporous & Mesoporous Materials
Catalysis Today	Reaction Kinetics & Catalysis Letters
Catalysis Letters	Studies in Surface Science & Catalysis

Table B.2. Set B: Title words alone indicative of catalysis.

Ag/Al_2O_3	Cocataly*	Pd/SiO_2
Au/Al_2O_3	$CoMo/Al_2O_3$	Photocataly*
Au/Fe_2O_3	Electrocataly*	Photoelectrocataly*
Au/SiO_2	Enzym* + Immobiliz*	Precatalyst*
Au/TiO_2	Fe*ZSm*	Pt/Al_2O_3*
Autocatal*	Nanocataly*	Pt/ZrO_2*
Biocatal*	Ni/Al_2O_3	Ru/Al_2O_3
Catalys*	Organocataly*	Sn/Al_2O_3
Catalyt*	Pd/Al_2O_3	Sonocatalysis

Table B.3. Set C: Title words indicative of catalytic activity.

Activat*	Desorption	Hydroformylat*	Oxidat*
Activit*	Desulfurizat*	Hydrogenat*	Oxygen Storage
Adsorb	DFT	Hydroisomerizat*	Photooxidat*
Adsorbate	Epoxidat*	Hydrolysis	Preparation
Adsorbates	Esterificat*	Hydroprocess*	Reaction*
Adsorbed	Friedel	Hydrotreat*	Reactivit*
Adsorbing	Growth	Hydroxylat*	Reduction*
Adsorbs	HDS	Immobiliz*	Reforming
Adsorption	Heck	Isomerizat*	Selectivit*
Adsorption/Desorption	Hydrocrack*	Kinetic*	Suzuki
Aromatizat*	Hydrodechlorinat*	Metathesis	Transesterificat*
Deactivat*	Hydrodemetalliz*	Methylat*	Trimerizat*
Dehydrat*	Hydrodenitrogenat*	Oligomerizat*	Tropsch
Dehydrogenat*			

Table B.4. Set D: Title words indicative of an element or structure used for catalysis.

Ag	Graphite	Molybdenum*	Re/*	Tin
Ag/*	$H_3PW_{12}O_{40}$*	Molybdophos*	Rhenium*	Tin/*
Au	$H_3PMo_{12}O_{40}$	Montmorillonite*	Rh	TiO_2*
Au/*	Heterogenous	Mordenite*	Rh/*	Titania
Carbide*	Heteropoly	Nanocluster*	Rhodium*	Titanium
Ce	Heteropolyacid*	Nanocrystalline	Ru	Tungsten
Ce/*	Hydrotalcite*	Nanoparticle*	Ru/*	Tungstophosphoric
CeO_2	Ir	Ni	Ruthenium*	V
Ceria	Ir/*	Ni/*	Salen	V/*
Cerium	Iridium*	Nickel*	Sapo	V_2O_5
Chromium*	Keggin	$NiMo/Al_2O_3$	SBA	Vanadia
Cluster*	La	Nitride*	Schiff	Vanadium*
Cobalt	La/*	Oxide*	Sieve*	Vanadyl
Co/*	Lanthanum	Palladium*	Silicalite	W
Copper	Manganese*	Pd	Sn	W/*
Cr	MCM	Pd/*	Sn/*	Zeolite*
Cr/*	Mesoporous	Perovskites	Sulfide*	Zirconia
Cu	Microporous	Pillared	Ta	Zirconium*
Cu/*	Mn	Platinum*	Ta/*	Zr
Fe	Mn/*	Polyoxometalate*	Tantalum	ZSM
Fe/*	Mo	Pt/*	Ti	
Gold*	Mo/*	Re	Ti/*	

Table B.5. Set E: Title words to exclude non-catalysis papers.

DNA*	Protein*	Protease*	RNA*	Solar Cell*

The filter was calibrated with reference to papers with and without address terms such as catalysis and its equivalents in some other European languages (although the titles of all papers are translated in the SCI into English, with US spelling, the words in the addresses are often left in the original language, transliterated as necessary into the Roman alphabet). Its precision was determined as 0.62 and its recall as 0.78. These are not particularly high values (values > 0.9 are not uncommon in biomedical subject areas), but then catalysis is a relatively diffuse subject. This means that the apparent world output of catalysis papers should be multiplied by 0.62/0.78 = 0.795 to obtain an estimate of the correct number of papers.

Table B.6. Title words used to characterize catalysis papers as being relevant to biomass as a source of renewable energy.

Bagasse	Biofuel OR Bio-Fuel
Biodiesel	Biomass NOT (Detox OR Sorption OR Removal)
Biooil OR Bio-Oil	Sugarcane OR Sugar-Cane
Glycerol AND (Hydrogenolysis OR Oxidation)	Sugar AND (Hydrolysis OR Oxidation)

During the study, attention was focused on the use of catalysis to promote the industrial exploitation of biomass for energy purposes, and it was decided to carry out a subanalysis of papers that appeared to be relevant to this subject area. They were selected by means of title words selected by one of the panelists, Prof. George Huber, and were as listed in Table B.6.

The addresses on the papers were analyzed by means of a special macro (written by Dr. Philip Roe) that determined for each paper the fractional count of each country listed in the addresses, and also the integer count. A paper with two addresses in the USA, one in France, and one in Germany would count 0.5 for the USA and 0.25 each for France and Germany on a fractional count basis; it would count unity for each country on an integer count basis. Fractional counts enable the totals for groupings of countries, such as the European Union or Latin America, to be determined; they are also more representative of a country's true contribution to the field.

On the other hand, integer counts enable a country's relative commitment (RC) to catalysis research to be compared with its presence in all science. This ratio shows the relative importance of catalysis research for the country, with the value for the world being unity. It has the advantage that it compensates, to some degree, for biases in the SCI for or against some countries because of language or other reasons, on the assumption that catalysis research papers are as likely as ones in other fields of science to be published in journals processed for the SCI.

The **potential citation impact** (PCI) of each of the papers in the file was determined on the assumption that they would be cited as frequently as other papers in the same journal. The citation time window was five years, i.e., the year of publication and four subsequent years. Data on the mean citation rates of papers in each journal were obtained from Thomson Scientific (the publishers of the SCI) for every other year in the period 1996–2002: they do not normally change much in a two-year period. The mean PCI for leading countries was then calculated for three cohorts of papers: 1996–1999, 2000–2002 and 2003–2005. Some values of PCI for leading journals used by catalysis researchers in the four periods are shown in Table B.7.

To meet the second objective of the study, namely to assist the panel in the identification of the leading institutions carrying out catalysis research in Europe and East Asia, it was necessary to perform an analysis of the institutional addresses for selected countries (three in East Asia and 13 in Europe). This is the first item in the address, and appears in contracted form, e.g., Chinese-Acad-Sci, Jilin-Univ, Natl-Inst-Mat-Sci. Sometimes different forms are used for institution names, and unification is needed in order to determine the true total for an institution. For the purposes of this study, a count was made of the numbers of addresses appearing during the last three years, viz., 2003–2005; this will usually be slightly more than the numbers of papers, but the ranking of institutions is unlikely to be much affected. Some national scientific institutions (e.g., Chinese-Acad-Sci, CNRS in France, CSIC in Spain, CNR in Italy) have

Table B.7. Leading journals used by catalysis researchers, their percentage usage in 2003–2005, and their potential citation impact factors in 1996–1997, 1998–1999, 2000–2001, and 2002+.

Journal	%	1996–	1998–	2000–	2002+
Angewandte Chemie-International Edition	1.28	20.3	21.2	23.2	29.3
Journal of the American Chemical Society	2.77	22.9	23.1	25.7	28.5
Applied Catalysis B-Environmental	1.59	12.4	14.5	16.7	20.2
Chemistry-a European Journal	0.62	18.3	18.3	18.3	17.2
Journal of Catalysis	2.71	12.9	12.2	14.1	16.0
Applied Catalysis A-General	4.05	8.2	7.4	9.4	14.9
Journal of Molecular Catalysis A-Chemical	3.93	6.4	7.3	6.4	10.2
Catalysis Letters	1.84	8.9	7.6	7.3	7.5
Reaction Kinetics and Catalysis Letters	0.99	1.8	1.9	2.1	2.4
Studies in Surface Science and Catalysis	2.82	3.5	2.3	2.3	1.9
Kinetics and Catalysis	0.72	0.6	0.7	0.7	1.1

B.4 Results: Quantitative Outputs

Figure B.1 shows the growth of catalysis research outputs compared with all science, as recorded in the SCI. (Allowance has been made for the shortfall of papers in 2005 because of late recording.) The former is growing at about 5.4% per year, compared with just 2.9% per year for all science, so it can be regarded as a rapidly-growing field. Figure B.2 shows the outputs of the leading countries, presented as three-year moving averages so as to smooth out annual variations, on a fractional count basis. "Europe 13" sums the output of the countries listed in Table B.8.

China's output has been growing rapidly, and in 2005 it was greater than that of Japan. The output of the 13 European countries listed in Table B.6 is collectively almost twice that of the United States; in most other fields of science, the outputs of Western Europe and the United States are similar in magnitude. The outputs of the leading European countries are shown in Figure B.3 on a different scale to that of Figure B.2.

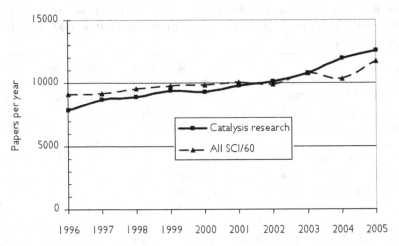

Fig. B.1. World outputs of catalysis research papers (after application of calibration factor), solid line, and of all science (divided by 60), dashed line, 1996–2005.

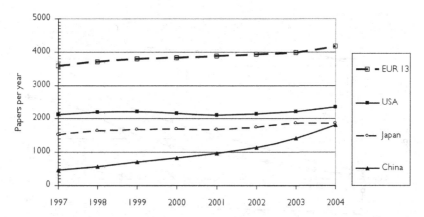

Fig. B.2. Outputs of catalysis research papers (not corrected for CF) from 13 European countries, the USA, Japan and China (People's Republic), 1996–2005, three-year averages.

Table B.8. 13 European countries used for institution location analysis.

Country	Code	Country	Code	Country	Code	Country	Code
Austria	AT	France	FR	Netherlands	NL	Sweden	SE
Belgium	BE	Germany	DE	Norway	NO	Switzerland	CH
Denmark	DK	Italy	IT	Spain	ES	UK	UK
Finland	FI						

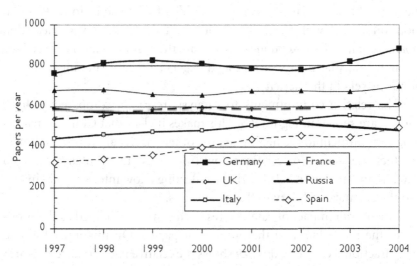

Fig. B.3. Catalysis research outputs (uncorrected) for six European countries, 1996–2005, three-year moving averages.

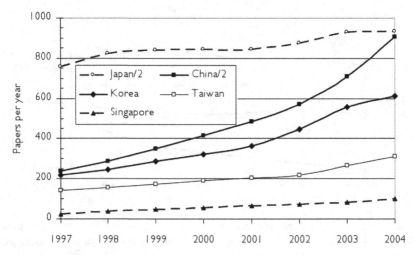

Fig. B.4. Catalysis research outputs (uncorrected) for five East Asian countries, 1996–2005, three-year moving averages.

Whereas the outputs of Germany, France, and the UK have been rather stable (although that of Germany is now increasing), and those of Italy and Spain have been increasing; that of Russia has been declining, and it is now the lowest of the six.

The outputs of some East Asian countries are shown in Figure B.4: those of China and Japan have been divided by two so as to make them more on a scale with those of the others. This makes it very clear that Japan, China, and the Republic of Korea are the main producers of catalysis research, and that the other so-called "tigers", Taiwan and Singapore, are minor players in this activity.

The relative commitments to catalysis research, compared with all science, are shown in Figure B.5. This makes it clear that China is making a big commitment to the field, as are the other East Asian countries shown, but that the field is of relatively low priority in the United States and Canada, and also in the UK, perhaps reflecting a low interest in chemistry and the chemical industry in these countries.

The overall impact of catalysis research can be calculated as the product of the mean PCI and the number of papers. On this indicator, and using fractional counts, Table B.9 shows the relative performance of North America (the USA + Canada), Europe 13 (with the contributions of the six

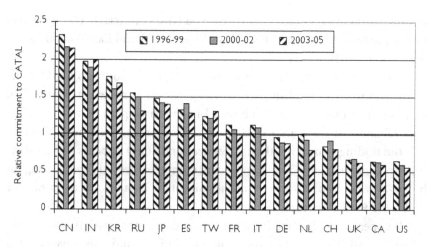

Fig. B.5. Relative commitments of leading countries to catalysis research, 1996–2005. CN = China, IN = India, KR = Korea, RU = Russia, JP = Japan, TW = Taiwan, CA = Canada.

Table B.9. Overall potential citation impact (PCI × number of papers) of catalysis research from different countries and world regions in 3 periods.

Region/Country	1996–1999	2000–2002	2003–2005
North America	120525	91001	95128
US	110464	82608	86164
CA	10060	8393	8965
Europe 13	126505	108063	111777
DE	27619	22476	23637
UK	20973	18030	17240
FR	21468	17158	16628
IT	14919	12689	14513
ES	11263	11498	13504
NL	8546	7270	6524
Others	21718	18942	19731
East Asia 5	68507	68715	90624
JP	45281	39143	41916
CN	12251	16235	29869
KR	5585	7887	10559
TW	4376	4071	6098
SG	1014	1380	2183
Ratio, Eur13/NA	1.05	1.19	1.18
Ratio, EA/NA	0.57	0.76	0.95

leading countries specified), and East Asia (five countries). The table also shows the ratio of the overall impact of Europe 13 and East Asia to North America in the three periods.

The mean potential citation impact (PCI) of the journals used by different countries to publish their catalysis papers is shown in Figure B.6. (The mean values for the whole world in the three periods were 8.4, 9.1, and 9.0 cites in five years.) Clearly on this criterion, the United States is still the leading country, although Switzerland (CH) appears to be catching up quite rapidly. The East Asian countries are still publishing in below-average impact factor journals, although Japan's output in not far below the world average.

It appears that Europe is increasing its dominance over that of North America, and that East Asia is catching up quickly and may have equaled the total impact of North America in 2007.

The output of catalysis papers relevant to biomass as a source of energy was small, only 233 in the 10-year period used for analysis, but it grew rapidly after 2001, as shown in Figure B.7.

Compared with all catalysis research, the USA published relatively less, and the 13 European countries relatively more, see Figure B.8.

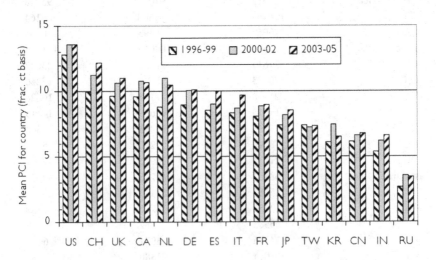

Fig. B.6. Mean Potential Citation Impact factor of journals used to publish catalysis research papers by leading countries, 1996–2005 (fractional count basis).

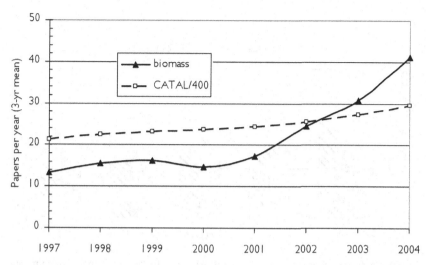

Fig. B.7. Output of catalysis papers (corrected for calibration factor and divided by 400) compared with that of papers relevant to biomass as a source of energy, 1996–2005, three-year running means.

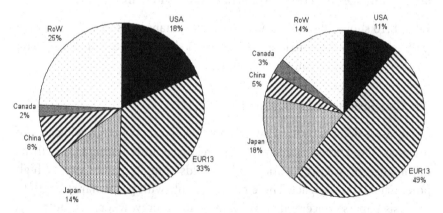

Fig. B.8. Geographical distribution of catalysis papers (*left*) and biomass ones (*right*).

However the United States published its relatively few biomass papers in high-impact journals, and in ones of higher PCI than those it used for catalysis papers overall, whereas the reverse was the case for all other countries; see Figure B.9.

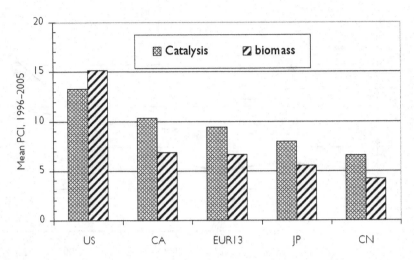

Fig. B.9. Mean five-year potential citation impact factors for biomass papers from different geographical areas, compared with those for all catalysis papers, 1996–2005.

B.5 Results: Leading Institutions in Catalysis Research

Table B.10 shows the leading institutions in catalysis research in the three leading East Asian countries, and Table B.11 shows the corresponding data for the 13 European countries listed in Table B.8.

B.6 Conclusions and Discussion

Catalysis, as defined by the evaluation panel, is a small but growing area of research. The United States publishes relatively little (compared with its dominance of science in most other fields) but its papers are of high potential citation impact. These remarks apply with particular strength to catalysis papers concerned with biomass for energy, where US output is only 11% of the world but the papers are in very high impact journals. European overall catalysis output is now almost twice that of the USA and some countries, notably Switzerland, are approaching it in terms of potential citation impact. In East Asia, China has now overtaken Japan in terms of output (this occurred in 2005), and Korea's publications are also increasing rapidly, though their potential citation impact is still quite low.

Table B.10. Leading catalysis research institutions in China, Japan, and Korea, 2003–2005 (numbers of addresses on catalysis papers).

Country	Institution	N
JP	Univ. Tokyo, Meguro Ku, Tokyo 1538904	996
JP	Tohoku Univ., Sendai, Miyagi 9808578	831
JP	Kyoto Univ., Kyoto 6068501	800
JP	Tokyo Inst. Technol., Yokohama, Kanagawa 226850	795
JP	Osaka Univ., Suita, Osaka 5650871	764
CN	Chinese Acad. Sci., Beijing 100080	748
JP	Hokkaido Univ., Sapporo, Hokkaido 001002	685
JP	Natl. Inst. Adv. Ind. Sci. & Technol., Tsukuba, 3058565, or Osaka 5638577	622
KR	Seoul Natl. Univ., Seoul 151742	622
CN	Chinese Acad. Sci., Shanghai 200032	544
CN	Dalian Inst. Chem. Phys., Dalian 116023	542
CN	Fudan Univ., Shanghai 200433	526
CN	Zhejiang Univ., Hangzhou 310032	510
CN	Tsing Hua Univ., Beijing 100084	484
JP	Kyushu Univ., Kasuga, Fukuoka 8168580	463
CN	Nanjing Univ., Nanjing 210093	454
CN	Jilin Univ., Changchun 130023	443
JP	Nagoya Univ., Nagoya, Aichi 4648603	435
KR	Korea Adv. Inst. Sci. & Technol., Taejon 305701	389
CN	Univ. Sci. & Technol. China, Hefei 230026, Anhui	387
CN	Xiamen Univ., Xiamen 361005	297
KR	Pohang Univ. Sci. & Technol., Pohang 790784	282
CN	Nankai Univ., Tianjin 300071	275
CN	Chinese Acad. Sci., Lanzhou 730000	272
JP	Sci. Univ. Tokyo, Shinjuku Ku, Tokyo 1628601	248
CN	Tianjin Univ., Tianjin 300072	237
KR	Hanyang Univ., Seoul 133791	236
CN	Hong Kong Polytech. Univ., Kowloon, Hong Kong	235

Some limitations of bibliometric analysis need to be pointed out. There are inevitable biases in the journal coverage of the SCI; in particular of journals not in English, which are bound to have a more limited readership. One effect of this is to *improve* the average PCI of papers from

Table B.11. Leading catalysis research institutions in 13 leading European countries, 2003–2005 (numbers of addresses on catalysis papers). For country codes, see Table B.8.

Country	Institution	N
ES	CSIC, 28040 Madrid	710
NL	Delft-Univ.-Technol.	450
IT	Univ.-Milan	436
FR	Univ.-Paris-06, 75252 Paris 05	416
UK	Univ.-Cambridge	405
FR	Inst.-Rech-Catalyse, 69626 Villeurbanne	380
CH	ETH-Honggerberg, 8093 Zurich	370
DE	Tech.-Univ.-Munich, 85747 Garching	363
IT	Univ.-Turin	332
NL	Univ.-Utrecht	301
NL	Eindhoven-Univ.-Technol.	300
DE	Max.-Planck-Gesell, Fritz Haber Inst, 14195 Berlin	297
FR	Univ.-Lyon-1, 69626 Villeurbanne	290
ES	Univ.-Barcelona, 08028 Barcelona	284
UK	Univ.-London-Imperial-Coll	283
CH	Ecole-Polytech.-Fed.-Lausanne, 1015 Lausanne	280
UK	Univ.-Liverpool	278
IT	Univ.-Roma-La-Sapienza	276
SE	Chalmers-Univ.-Technol., 41296 Gothenburg	271
BE	Univ.-Catholique-Louvain, 1348 Louvain	267
ES	CSIC, UPV, Inst. Technol. Quim., 46022 Valencia	264
IT	Univ.-Padua	259
FR	Univ.-Strasbourg-1, 67070 Strasbourg	243
UK	Univ.-Oxford	236
DE	Max.-Planck-Inst.-Kohlenforsch, 45470 Mülheim	233
UK	Cardiff-Univ.	207
DK	Tech-Univ.-Denmark, 2800 Lyngby	172
DK	Haldor-Topsøe-Res-Labs, 2800 Lyngby	109

Note: The last three institutions are ones visited by the panel, but they are not the next three in rank order. Not all 13 countries have an institution with 230 or more addresses in 2003–2005.

countries whose output has been affected, as the papers in purely national journals, with low citation counts, have not been taken into consideration. However it is likely that the best labs in these countries will publish the large majority of their papers in international journals.

Another limitation is that counts of papers in the SCI may not reflect the amount of research activity in commercial companies, where considerations of confidentiality will inevitably limit the amount that they seek to publish. This is likely to be of more importance in a field such as chemistry than in biomedicine, where publication of clinical trial results is a pre-requisite to the licensing of a new drug, and publication of new research achievements is the way to advancement for young biotech companies. Thus Shell in Amsterdam, a major company in catalysis, only published 11 papers in 2003–2005.

No attempt has been made to determine the actual citation impact of papers from different countries as a check on the values of potential impact based on journals. This was primarily because the number of papers was so large (more than 125,000 over the decade) that even a sample of several thousand might have been quite unrepresentative. Citation counts are but one measure of "quality". The links between universities and industry may be a more meaningful indicator of the relevance of the research being carried out in academia. These could be co-authorships, but other, less visible links such as research sponsorship, paid consultancies and industrial placement of students, could also be indicative of industrial relevance.

Appendix 3: Glossary

^{1}H NMR	Proton nuclear magnetic resonance
ADF	Annular dark-field
AES	Auger electron spectroscopy
AFM	Atomic force microscope
AIST	National Institute of Advanced Industrial Science and Technology (Japan)
ATR-IR	Attenuated total reflectance infrared (spectroscopy)
Au	Gold (chemical symbol)
BCC	Biomass catalytic cracking
BESSY (II)	Berliner Elektronenspeicherring-Gesellschaft für Synchrotronstrahlung (3rd-generation synchrotron radiation facility)
BET	Brunauer–Emmett–Teller surface area characterization/structure analysis method
BK21	Brain Korea 21st Century program
BN	Boron nitride
BOE	Barrels of oil equivalent

BSC-CNS	Barcelona Supercomputing Center — Centro Nacional de Supercomputación (the National Supercomputing Facility and National Supercomputing Center of Spain)
CAS	Chinese Academy of Sciences
CESCA	Centre de Supercomputació de Catalunya (Spain)
CESGA	Centro de Supercomputación de Galicia (Spain)
CNF	Carbon nanofiber(s)
CNR	Consiglio Nazionale delle Ricerche (National Research Council of Italy)
CNRS	Centre National de la Recherche Scientifique (National Center for Scientific Research, under the Ministry of Research, France)
CNTs	Carbon nanotubes
CO	Carbon monoxide (chemical symbol)
CPU	Central processing unit (computer)
CREST	Core Research for Evolutional Science and Technology (Japan Science and Technology Agency and the Research Development Corporation of Japan)
Cs	Cesium (chemical symbol)
CSIC	Consejo Superior de Investigaciones Cientificas (Council for Scientific Research, Spain)
CTI	Centro Técnico de Informática (Spain)
CUPS	KAIST's Center for Ultra-microchemical Process Systems
CVD	Chemical vapor deposition
DABCO	Diazabicyclo (2,2) octane
DCC	Deep catalytic cracking
DEFC	Direct ethanol fuel cell
DEMS	Differential electrochemical mass spectrometry
DFT	Density functional theory
DFTB	Density functional tight-binding
DICP	Dalian Institute of Chemical Physics, Chinese Academy of Sciences
DME	Dimethyl ether, a new fuel with properties similar to LPG (liquefied petroleum gas, propane gas), expected to replace diesel fuel and LPG

DMFC	Direct methanol fuel cell
DRIFTS	Diffuse reflectance infrared Fourier transform spectroscopy
DSC	Differential scanning calorimetry
DTG	Derivative thermogravimetry
DTU	Technical University of Denmark
EAC	Structure and Activity of Catalysts Group at the Institute of Catalysis and Petrochemistry (Spain)
EB	Ethylbenzene
EDS	Energy dispersive spectroscopy
EDX	Energy dispersive X-ray (micro)analysis/spectroscopy
ee	Enantiomeric excess
EELS	Electron energy-loss spectroscopy
ELCASS	European Laboratory (Consortium) of Catalysis and Surface Science
EMR	Electron magnetic resonance
EPR	Electron paramagnetic resonance, also known as electron spin resonance (ESR)
EPSRC	Engineering and Physical Sciences Research ouncil (UK)
ESEM	Environmental scanning electron microscope/y
ESR	Equivalent series resistance/resistant (measurement, calculation); electron spin resonance
ESRF	European Synchrotron Radiation Facility (Grenoble, France)
ETEM	Environmental transmission electron microscopy
EtOH	Ethanol
EUBIA	European Biomass Industry Association
EXAFS	Extended X-ray absorption fine structure
EXPEEM	Energy-filtered X-ray photoemission electron microscopy
FCC	Fluid catalytic cracking
fcc	Face centered cubic (structure)
FEG	Field-emission gun
FEI	FEI Company (multinational/German)
FESEM	Field-emission SEM
FHI	Fritz Haber Institute (Germany)

FIB	Focused ion beam
FID	Flame ionization detection
FP	Flame pyrolysis
FTIR	Fourier transform infrared (spectroscopy)
FTS	Fischer–Tropsch synthesis
GC	Gas chromatography
GIF	Gatan imaging filter
GTL	Gas to liquid
HAADF	High-angle annular dark-field
hcp	Hexagonal closest packing (structure)
HDM	Hydrodemetallization
HDO	Hydrodeoxygenation
HDS	Hydrodesulfurization
HMS	Hexagonal mesoporous silica
HPA	Heteropolyanions
HPLC	High-pressure liquid chromatography; high-performance liquid chromatography
HP-RAIRS	High-pressure reflection-absorption infrared spectroscopy
HREELS	High-resolution EELS
HREM	High-resolution electron microscopy
HRTEM	High-resolution transmission electron microscopy
ICP	Institute of Catalysis and Petrochemistry (Spain)
IDECAT	Integrated Design of Catalytic Nanomaterials for Sustainable Production (EU)
INSTM	Interuniversity Consortium on Science and Technology of Materials (Italy)
IP	Intellectual property
IR	Infrared
ISS	Ion scattering spectroscopy
ITQ	Instituto de Tecnología Química (Spain)
JSPS	Japan Society for the Promotion of Science
JST	Japan Science and Technology Agency
KAIST	Korea Advanced Institute of Science and Technology
KEK-PF	High Energy Accelerator Research Organization of Japan, Photon Factory

LEED	Low-energy electron diffraction
LEEM	Low-energy electron microscopy
LEIS	Low-energy ion scattering
LISF	Laser-induced surface fluorescence spectroscopy
LT-STM	Low-temperature scanning tunneling microscopy
MAS	Magic angle spinning
MCH	Methylcyclohexane
MeOH	Methanol
MEXT	Ministry of Education, Culture, Sports, Science and Technology (Japan)
MITI	Ministry of International Trade and Industry (Japan)
MRI	Magnetic resonance imaging
MS	Mass spectrometry
MTG	Methane to gasoline
MTO	Methanol to olefins
MWW	A type of zeolite structure
NEDO	New Energy and Industrial Technology Development Organization (Japan)
NEMAS	NanoEngineering Materials and Surfaces Center of the Politecnico di Milano (Italy)
NER	Nanoscale Exploratory Research programs of NSF
NEXAFS	Near-edge X-ray absorption fine structure
NIOK	Netherlands Institute for Catalysis Research
NIR	Near infrared
NIRT	Nanoscale Interdisciplinary Research Teams programs of NSF
NMR	Nuclear magnetic resonance
NoE	European Network of Excellence
NRSCC	National Research School of Combination Catalysis (The Netherlands)
NSF	National Science Foundation (U.S.)
ODH	Oxidative dehydrogenation reactions
OIM	Orientation image microscopy
OSC	Oxygen storage capacity
PAH	Polycyclic aromatic hydrocarbon
PDADMAC	Polydiallyldimethylammonium chloride

PED	Photoelectron diffraction
PEEM	Photoemission electron microscopy
PEM	Polymer electrolyte membrane (fuel cell)
PI	Principal investigator
PIRE	Partnership for International Research and Education program of NSF
PM	Particulate matter
PM-IRRAS	Polarization modulation-infrared reflection-absorption spectroscopy
PMO	Periodic mesoporous organosilica
PO	Propylene oxide
PrOx	Preferential oxidation (also PROX)
PSOFC	Pressurized solid oxide fuel cell
PTRF or PTRF-EXAFS	Polarization-dependent total-reflection fluorescence EXAFS
PVP	Poly(N-vinylpyrrolidone)
QEXAFS	Quick extended X-ray absorption fine structure
QM/MM	Quantum mechanical/molecular mechanical
QMS	Quadrupole mass spectrometer
QXAFS	Quick X-ray absorption fine structure
RAIRS	Reflection-absorption infrared spectroscopy
RC	Relative commitment of a country (to catalysis research measured in the bibliometric study, Appendix B)
RMB	Ren min bi, national currency of China
SAXS	Small-angle X-ray scattering
SBA-15	Mesoporous silica
SCAC	Single-crystal adsorption calorimetry
SCR	Selective catalytic reduction
SEM	Scanning electron microscopy/microscope
SFG	Sum frequency generation
SIMS	Secondary ion mass spectrometry (also, dynamic SIMS)
SKLC	State Key Laboratories of Catalysis (DICP, China)
SKLMRD	State Key Laboratory of Molecular Reaction Dynamics (DICP, China)

SMART	Ultrahigh-resolution spectromicroscope (SpectroMicroscope for All Relevant Techniques) (FHI, for use in BESSY)
SOFC	Solid oxide fuel cell
SPring-8	Large 3rd-generation synchrotron radiation facility, Harima Science Garden City Hyogo, Japan, of the Japan Synchrotron Radiation Research Institute (JASRI)
SSNMR	Solid-state NMR spectroscopy
STEM	Scanning transmission electron microscopy/microscope
STM/STS	Scanning tunneling microscopy/spectroscopy
SWCNT	Single-walled carbon nanotubes
syngas	Synthesis gas, a gas mixture that contains varying amounts of carbon monoxide and hydrogen generated by the gasification of a carbon-containing fuel to a gaseous product with a heating value
TAP	Temporal analysis of products
TBP	Tributylphosphine
TBPP	Tetra-3,5-di-ter-butyl-phenyl porphyrin
TDS	Thermal desorption spectroscopy
TEAOH	Tetraethylammonium hydroxide
TEDDI	Tomographic energy dispersive diffraction imaging
TEM	Transmission electron microscopy/microscope
TGA	Thermogravimetric analysis
TGA-DTA	Thermal gravimetric (or thermogravimetric) analysis — differential thermal analysis
TMCH	3,3,5 trimethyl cyclohexanone
tmeda	Tetramethylethylenediamine(also TMEDA)
TMP	Trimethylpentane isomers
TNT	Titania nanostructured thin films
TOF	Turnover frequency/ies
TPD	Temperature-programmed desorption
TPR/TPO	Temperature-programmed reduction/oxidation
TWC	Three-way catalysts or three-way catalytic converter for auto exhaust

UHV	Ultrahigh-vacuum
UHV-RAIRS	Ultrahigh-vacuum reflection-absorption infrared spectroscopy
ULSD	Ultra-low-sulfur diesel
UPS	Ultraviolet photoelectron spectroscopy
USY	Ultrastable Y (zeolite)
UV	Ultraviolet
UV/Vis	Ultraviolet-visible (spectroscopy or spectrophotometry)
vis	Visible (spectrum, spectroscopy)
VOC	Volatile organic compound
VPO	Vanadium phosphorus oxide
VUV	Vacuum Ultraviolet
WAXS	Wide-angle X-ray scattering
WC	Tungsten carbide
WDS	Wavelength-dispersive spectrometer/try
WGS	Water gas shift (reactor, catalyst, etc.)
WHSV	Weight hourly space velocity
XAFS	X-ray absorption fine structure
XANAM	X-ray-aided non-contact atomic force microscopy
XANES	X-ray absorption near-edge structure
XAS	X-ray absorption spectroscopy
Xe	Xenon (chemical symbol)
XEDS	X-ray energy dispersive spectroscopy
XPEEM	X-ray photoemission electron microscopy
XPS	X-ray photoelectron spectroscopy
XRD	X-ray diffraction
ZP	Zeolite precursors

Index

absorbents 153
activation barriers 97
Air Force Office of Scientific Research (AFOSR) v
alkylation, dehydrogenation, hydrogenation 186
alumina 74
aluminosilicates 34
ammonia 28
aqueous-phase processing 251
atomic force microscopy 66
Attenuated Total Reflectance (ATR) 47
Auger spectroscopy 52

benzaldehyde 33
benzene hydroxylation 28
biocatalysis xxiii
bioenergy 16
biomass xxv, 241

biorenewables 130

Canonical Grand Canonical 97
carbon 74
carbon nanotubes (CNTs) 84
Catalysis and Biocatalysis Program xx
catalytic activity 3
catalytic hydrodesulfurization (HDS) 153
catalytic hydrogenation 2
cellulose 251
ceria 84
characterization xxv, 4
China xxv
CO_2 16
coal xxv
coal conversion 130
coke 49

computational catalysis 94
configurational interactions (CI) 95

Defense Threat Reduction Agency (DTRA) v
density functional theory (DFT) 96, 97, 156
Department of Energy (DOE) v
diesel fuel 251

economic impact of catalysis 1
ECR-34 37
electrocatalysis 130
electron diffraction 83
electron microscopy 83
electron paramagnetic resonance (EPR) 44
electron paramagnetic resonance spectrometer 51
electron spectroscopies 44
environmental
environmental SEM 87
ethanol 251
extended X-ray absorption fine structure, or EXAFS 55

feedstock 241
Fe-ZSM-5 48
Fischer–Tropsch synthesis 84, 130
FTS 245
fuel cells 16

gas chromatography, physisorption and chemisorption diffraction (XRD) 44
gasification 245, 251
gasoline 251

Gibbs Ensemble Monte Carlo simulations 97
Gibbs–Thompson effect 76
Grand Canonical Ensemble Monte Carlo (GCMC) simulations 97

Hartree–Fock (HF) 95
hemi-cellulose 251
heptanaldehyde 33
heterogeneous catalysis 2, 25
heteropolyanions (HPA) 39
hexagonal mesoporous silica (HMS) 157
hexamethonium 37
highangle annular dark-field (HAADF) detector 69
high-resolution TEM 67, 83
high-throughput 21
homogeneous 2
homogeneous catalysts 92
homogeneous or single-site molecular catalysis xxiii
Horiuchi–Polanyi mechanism 206
hydrocarbon 2
hydrogen 16
hydrogen peroxide 72
hydrogenation, hydrocarbon hydrocracking 84

IR spectroscopy 48
ITQ-21 195
ITQ-33 37

jasminaldehyde 33

Korea xxiv

LEIS 228
lignin 251

liquefaction 251
liquid-phase catalytic processing 251
low-energy electron diffraction (LEED) 52
low-energy electron microscopy (LEEM) 52, 70

mesoporosity 32
mesostructure 32
metal organic frameworks (MOF) 205
methane conversion 130
microkinetic 100
Mo 48
model surfaces xxiii
modeling 4
morphology 32, 65

nanoparticles 186
Nanoscale Exploratory Research (NER) xx
Nanoscale Interdisciplinary Research Teams (NIRT) xx
National Aeronautics and Space Administration (NASA) 172
National Science Foundation (NSF) v
natural gas xxv
NEXAFS (near edge X-ray absorption fine structure) 52, 118
nitrogen 29
NMR imaging (MRI) 48
nuclear magnetic resonance (NMR) 44

Office of Science and Technology Policy (OSTP) xiii, xiv
Ostwald ripening 76

oxidation 2, 30
oxygen 29

Partnerships for International Research and Education (PIRE) xx
pathways 94
periodic mesoporous organosilicas (PMOs) 187
petrochemical 84
phenol 28
photo emission electron microscopy 70
photocatalysis 130
photocatalytic water splitting 256
photoelectron 52
photoelectron diffraction (PED) 52
photoelectron emission microscopes (PEEM) 51
platinum (Pt) 26
polydiallyldimethylammonium chloride (PDADMAC) 190
polyhedral oligomeric silsesquioxanes (POSS) 200
polymer electrolyte membrane fuel cells (PEMFC) 167
polyol process 210
pyrolysis 251

Quantum Monte Carlo (QMC) 95

Raman, and infrared (IR) 44
reactivity 65
redox 84
rhenium 28
rhenium oxynitride 28
rhodium 3

scanning electron microscopy (SEM) 44, 67
scanning probe microscopy 66
scanning tunneling microscopy (STM) 51, 66
scanning-transmission electron microscopy 67
selective oxidation 186
selectivity 3
sinter 76
solid oxide fuel cells (SOFC) 167
Spain xxiv
spectroscopy xxv, 44
stochastic kinetic approach 101
synchrotron radiation 52
synthesis 4

tert-Bu hydrogen peroxide (TBHP) 200
tetraethylammonium hydroxide (TEAOH) 190
tetraethylorthosilicate 36
tetrapropylammonium 36
titania 71
titania nanostructured thin films 71
titanium silsesquioxane (TiPOSS) complex 200
transition state theory 97
transition states 97
transmission electron microscope 68

transmission electron microscopy (TEM) 44
transport 16
trimethylpentane isomers (TMP) 195
trioxomethylrhenium 28

ultrahigh-vacuum (UHV) 52
ultraviolet/visible (UV/Vis) 44
USY 39

variational transition state theory 97
vegetable oils 252

water gas shift (WGS) reaction 158
wave function methods 97
World Technology Evaluation Center (WTEC) xiv

XPS (X-ray photoelectron) 52, 228
X-ray absorption near-edge structure, or XANES 55
X-ray photoemission electron microscopy (XPEEM) 52

zeolite xxiv
Ziegler–Natta catalysts 52
zinc oxide (ZnO) 153
ZPITQ-21 195
ZSM-12 195
ZSM-5 32